The Capitalist Imperative

For Nadya and Suzanna
(Paulo and Trevor, too)

THE CAPITALIST IMPERATIVE

Territory, Technology, and Industrial Growth

Michael Storper and Richard Walker

Basil Blackwell

First published 1989

Basil Blackwell Inc.
432 Park Avenue South, Suite 1503
New York, NY 10016, USA

Basil Blackwell Ltd
108 Cowley Road, Oxford, OX4 1JF, UK

Library of Congress Cataloging in Publication Data

Storper, Michael.
The capitalist imperative.

Bibliography: p.
Includes index.
1. Industry – Location. 2. Capitalism.
3. Technological innovations – Economic aspects.
4. Economic development. I. Walker, Richard,
1947– . II. Title.
HD58.S6786 1989 338 89-900
ISBN 0-631-15625-9
ISBN 0-631-16533-9 (pbk.)

British Library Cataloguing in Publication Data

A CIP catalogue record for this book is available from the British Library.

Phototypeset in 10 on 11½ pt Sabon
by Dobbie Typesetting Limited, Plymouth, Devon
Printed in Great Britain by TJ Press Ltd., Padstow, Cornwall.

Contents

The cities that were formerly great have most of them become insignificant; and such as are at present powerful, were weak in olden time. I shall therefore discourse equally of both, convinced that human happiness never continues long in one stay.

Herodotus

Preface

This book, like all human projects, grew from a particular material base set in time and place. Our good fortune has been to live and work in California. This amazing state is now to be numbered among the mightiest industrial economies on earth, and serves as the core territory for the aerospace, microelectronics and film industries, a *troika* of powers for good and ill like none other in today's world. Those living in the eastern United States and in Europe are, it must be said, often disinclined to take Californians seriously; the latter reciprocate by exporting the personal computer, Star Wars, and Ronald Reagan. Our perspective has been skewed in important ways by the view from California, on the edge of the booming northern Pacific Rim; hence the overriding emphasis given here to economic growth and geographic expansion in modern industrialization. To the observer looking out from the brutal terrain of Liverpool's docklands, south Chicago, or the slums of Kingston, this emphasis will doubtless appear to give a rosy tint to capitalism's forward march, painting a picture that does little justice to those left behind, left out, or ground beneath the juggernaut. Nevertheless, others are better positioned to depict the devastation of unemployment in the First World or underdevelopment and imperialism in the Third.

We have also been favored in time by an efflorescence in Anglo-American social theory, a revival of the academic left in the United States, and especially by a thorough-going restructuring of economic geography by a gifted generation of radical thinkers. We have benefited immeasurably by our association – as students, colleagues and friends – with Bennett Harrison, David Harvey, Alain Lipietz, Doreen Massey, and Andrew Sayer, who have played leading roles in the difficult but rewarding task of moving geography, regional theory, and urban studies out of the intellectual ghetto to which they have been too long consigned. We have shared in this process of creating a new economic geography with a marvelous cohort that includes Phil Cooke, Gordon Clark, Ruth Fincher, Richard Florida, Derek Gregory, Kevin Morgan, Dick Peet, Eric Sheppard, Neil Smith, and Michael Webber, to mention just a few. Thanks, too, to those outside the field who have come to appreciate the spatial dimension in human existence, including Michael Burawoy, Manuel Castells, Mike Davis, Anthony Giddens, Charles Sabel, and John Urry.

At a narrower compass, we have had the extreme good geographic timing to have crossed paths in Berkeley and Los Angeles with an extraordinary number of gifted people of commensurate purpose and congenial presence. In this emergent "California School" of geographers and planners we number not only those still residing in the core region, but also many now dispersed to far-flung growth peripheries. Our deep thanks to Judy Carney, Susan Christopherson, Michael Dear, Marshall Feldman, Margaret Fitzsimmons, John Friedmann, Meric Gertler, Amy Glasmeier, Doug Greenberg, Peter Hall, Susanna Hecht, Michael Heiman, Ann Markusen, Rebecca Morales, Kristin Nelson, Allan Pred, Mary Beth Pudup, Anno Saxenian, Erica Schoenberger, Allen Scott, Phil Shapira, Ed Soja, Mike Teitz, Michael Watts, Marc Weiss, David Wilmoth, and Jennifer Wolch. Such a constellation of minds and friendships comes along rarely. We also extend our appreciation to contemporary students, soon to join the diaspora, such as Josh Muldavin, Brian Page, Arantxa Rodriguez, Tim Sturgeon, and Liz Vasile. In Los Angeles, students in a series of research seminars on technological change and growth centers have aided immeasurably in developing ideas presented here.

Finally, there is a mutually rewarding collaboration and friendship that extends back over a decade. We have struggled for several years, together and separately, to make sense out of industrialization, and a synthetic statement has at last been possible. This book is in every way a joint effort. Authorship is listed in alphabetical order: we discourage anyone from according primacy to either of us based on age, height, or the title page, or from trying to ascribe to either one of us particular ideas, however helpful or unfruitful they turn out to be. Chapters 1, 3, 7, and 8 were originally drafted by Michael Storper and chapters 2, 4, and 5 by Dick Walker; chapter 6 derives from an earlier paper, much amended. Everything has been worked over several times by us both, and we believe the final product is greater than the sum of the parts. We have had strenuous differences, of course, on such critical issues as value theory, structuralism, and post-Fordism, but we have tried to forego doctrinal fencing when there was a plain case to be made. Despite certain fundamental theoretical differences, it has been possible to come to workable positions on the basis of those broad principles on which we agree.

Before placing all credit at the tip of the academic edifice, let us recognize our sometimes invisible means of support. We would like to take this opportunity to thank the Geography and Regional Science Program of the National Science Foundation, the John Randolph Haynes and Dora Haynes Foundation of Los Angeles, the Fulbright Western European Research Program, and both branches of the University of California for research support in the past that contributed to this effort. Special thanks must be accorded to Peter Solomon for the clarity introduced by his patient editing and for his steadying influence in the face of numerous revisions;

to Charlie Hadenfeldt for his tendentious typing and ineffable good humor (not to mention his two-handed racquetball play); to Chic Dabby for her sharp-eyed proofreading; to Cherie Semans and Adrienne Morgan for their cartographic assistance; and to Doty Valrey, Natalia Vonnegut, Jaleezah Eskew, and Suzanne Masica in Berkeley for their help in typing and in many other ways. In Los Angeles, the staff of the Graduate School of Architecture and Urban Planning are all to be thanked for their assistance. We are also greatly indebted to John Davey of Basil Blackwell for showing confidence in us.

Our great loves, Chic Dabby and Susanna Hecht, figure in everything we do, not least of all in the writing of this book. Nadya Dabby deserves considerable solace for the time lost to a book project she has tried gamely to understand, but does not altogether appreciate, for good reason. May her future be as brilliant as her first twelve years. Credit is certainly due our parents – Bob and Louise Walker and Marilyn Munson – for their manifestly sound efforts on our behalf. And to Ellen Widess, a marvelous comrade who taught us both so much about the real political economy, a special word of thanks for times past.

Berkeley and Los Angeles, November 1988

Introduction

This book takes up two theoretical problems which exist in a mirror-image relationship to each other. On one hand, it is concerned with building an analytical framework for understanding the geography of economic development, specifically that of industrialization. On the other, it is at pains to show that political economic processes in general are profoundly shaped by their geography, and that any theoretical apparatus in the social sciences which ignores the geographical dimensions of these processes (as nearly all do in the twentieth century), does so at its own peril. These are grand claims and grand projects, and this book makes only a modest contribution to each. We believe that contribution is particularly crisp, however, because it looks at these broader issues through a large and clear window: industrialization.

Economic development is highly variable from place to place, and there seems to be no trend toward the evening-out of urban and regional economic patterns or of the fortunes of people in different territories. The optimism of modernization theories as late as the 1960s – with their predictions that jet planes, super-highways, and telecommunications would eliminate frictions of distance and lead us toward global modernization and equilibrium – is now dashed. At the same time, spectacular development has been unleashed in places where such positivist theories once claimed it was definitionally impossible. It is precisely the variability and volatility of regional fortunes at all scales (whether the neighborhood, city, subnational region, country, or subcontinent) that have become great issues of the 1980s, as territorially-defined political economies struggle with how to survive and improve their fates in the face of shifting rules and changing strategies of economic and political competition.

Curiously enough, the specialized domain known as industrial location theory has not figured prominently in these debates and struggles. More striking still, the term "industrial location" is virtually absent from the enormous literature on these pressing issues of economic development. All manner of discussion of development, employment, technology transfer, urbanization and trade – even that which recognizes industrialization as the major force in shaping these processes – goes on without reference to the simultaneous development of industries and their locations.

In part, this curious state of affairs is a consequence of the academic division of labor. Most of the debate about development is carried out in fields other than industrial location analysis: international economics, science policy research, labor studies, regional planning, the sociology of industrial societies, and so forth. In other words the geographical subject is implicitly at the center of major debates, but a specialized geographical theory is not. This naturally demands redress by geographers.

The situation becomes somewhat less curious in light of academic history. Early in the twentieth century, the social sciences lost their taste for *space* and increasingly focused on the role of *time* in human affairs; as the disciplines took on their modern forms, inherent historical specificity of place did not fit their intellectual priorities, which were to explain and often defend modernization (Gregory, 1978; Soja, 1988). Thus, the neglect of space is more likely a cause than a consequence of the academic division of labor. The political priorities of modern social science, especially in America, influenced the academic division of labor and helped make thought about the spatiality of social life a marginal concern. This creates a self-fulfilling prophecy. The body of specialized theory devoted to industrial location (indeed, to geographic problems in general) has become largely derivative of other social sciences.

Of course, each body of social theory must build on others, as it grapples with problems as complex as economic growth, and there are always complementary theories operating behind any explanatory framework. But location theory and economic geography have generally borrowed aspatial theories from other fields and simply extended them in such obvious ways that their geographical applications are rendered uninteresting. Space is introduced by referring to distance-sensitive or geographically-differentiated cost functions, or by accounting for institutional peculiarities in terms of territorial boundaries. Thus, the locus of explanation for industrialization and industrial location now lies outside location theory itself, and the real theoretical work goes on elsewhere. Industrial location theory has become little more than an extension of general equilibrium theory, organization theory or systems theory.[1] In spite of the empirical richness and diversity of regional economic histories, and the rapid pace of change in regional fortunes, industrial location theory took shape as simply the geographical extension of essentially static, mechanistic theories and models in economics, politics and sociology. So constrained, it gave up what might have been interesting about its subject and became incapable of making its own explanations. It was left to devise ever more complicated and arcane models of transport surfaces, while the theories that drove the models were developed in other disciplines. This sort of spatiality in social

[1] At the level of "pure" theory, only "urban economics" has managed to carve out a distinctive theoretical niche within neoclassical theory, as a result of the distinctive problems posed by land.

science is not the answer. The appellation "location theory" says it all: activities, once formed, must thereafter find a place to locate. The geography of industrialization was reduced to the study of the allocation of industrial plants across the economic landscape.

A necessary response to this state of affairs must therefore be to eliminate poor background theories which have ejected space and time and to adopt theoretical frameworks in which a geographical social science can be advanced. Neoclassical economic models, and their epistemological counterparts in the other disciplines, need to be replaced at the foundation. We therefore prefer Marx to Smith, Schumpeter to Marshall, Keynes to Walras, and Mills to Weber. We add the modifier "political" to economics, "historical" to sociology, and so bring growth and change to the forefront. We reject statics and allocation in favor of development and dynamics. With this, space must again become an active variable in the social system, because human action takes place in specific locales and social relations form both within and across territorial boundaries. The absolute and relational dimensions of space must be given priority over the prevailing idea of relative space as a matter of points and distances (Harvey, 1973). That is, human life unfolds in a thoroughly geographical way that is not contingent on other, more fundamental, dimensions of social process but is completely imbricated in both the structure and agency of history (Gregory and Urry, 1985). The consequence of acknowledging the essential spatiality as well as temporality of existence is a recentering of geographic inquiry within the domain of academic social science and the creation of a geographical historical materialism (Harvey, 1984).

We propose to thoroughly rewrite location theory from the standpoint of political economy, and, in so doing, to leave location theory behind in favor of a theory of geographical and territorial industrialization. In this effort, we draw on the work of a group of scholars who have been studying these problems for more than a decade. One task of this book is to pull together several of the pieces of this revolution in economic geography, in order to furnish a systematic picture of theoretical advances, while undergirding it with an integrative framework involving a number of new elements. Our primary aim is to move toward a synthesis to replace the neoclassical one locked into position by Isard (1956, 1969) and never fully supplanted by such challengers as behaviorism or systems theory. A second task is to push the new economic geography beyond the rather rigid conception of "spatial divisions of labor" that now prevails. Models based on spatial hierarchies of management, labor skills, or capital flows, made popular on the left in the 1970s by Hymer (1972), Holland (1976), and Froebel, Heinrichs and Kreye (1977), lack a sufficiently dynamic sense of unevenness and instability in the space-economy. Even Massey's (1984) pioneering work, which set the agenda for economic geography in the 1980s, can now be seen as having one foot still firmly planted in the previous decade. "Restructuring" theory, for all the insight it allows into processes of industrial change (Massey and Meegan, 1978, 1982), amounts to no more than comparative statics unless it is encompassed within a suitable theory of capitalist growth. Industry and corporate studies have proved enormously useful in deepening our understanding of capitalist production, but they are no substitute for economic theory.

There are four angles to the geometry of this theoretical edifice, and thus four ways this book can be read. The first is to theorize industry location as a process of industrial development in a spatial context, or "geographical industrialization." The various chapters lay out the bare bones of capitalist industrialization: the disequilibrium forces propelling industrial growth; the production of industrial places; the divergent paths and geographical specificity of technological development among industries; the integration of the division of labor into systems of production, particularly industrial and territorial complexes; and labor markets and the politics of employment in workplaces and communities.

The second way of reading this book is in terms of the role of places in the dynamics of capitalism, or what may be called "territorial development." This brings together the locational patterns of geographical industrialization at a more macroeconomic and macrogeographic level. Capitalist growth takes place through sequences of industries and industrial revolutions, in which technologically-advanced ensembles of sectors, based around product groups and associated production technologies, play a key part. We examine these technological dynamics and the way new heartland technologies can transform the productive forces of entire economies from time to time, by permitting the redesign of their most basic capital and consumer goods. We trace the problems of organizing interrelated and shifting production systems, and the dynamic possibilities opened up by territorial production clusters. We consider the effect of industrial change on the volatility of employment relations and the balance of power between workers and capitalists. We then look at how groups of industries effect radical changes in the overall geographical arrangement of economic activity, while renewing the bases of capitalist growth. Dynamic ensembles frequently move outside existing dominant territories to establish new growth regions, thereby expanding the working territory of capitalist society and elevating new cities to prominence, while reshaping older cities and industrial regions. At the same time, these territorial shifts shape class relations as a whole by their effect on income distribution, political power, and the conditions of class formation.

A third task of this book is the attempt to unify geographical and territorial industrialization – topics once treated separately as "industrial location" and "regional development" – in a single exploratory framework for viewing the macrogeography of capitalist economies. Throughout, we are at pains to account for three main features of the uneven development of capitalism: territorial expansionism, continuing differentiation of places, and instability in the relations between places. The geography of capitalism is uneven, to be sure; but it is, above all, inconstant. The dynamism of capitalist growth keeps industries on the move, and periodically sends them hurtling down new paths of spatial development. Technology, organization and labor all contribute to the systematic differences in industry growth paths, and the locational patterns that follow from those differential movements. Furthermore, each wave of industrialization brings into existence new growth centers and growth peripheries, stimulates disinvestment in some areas and the radical restructuring of others, and reshuffles spatial production relations and patterns of territorial income distribution and politics. In so doing, it gives new life to capitalism. Thus, locational particulars matter; yet capitalism's geographical processes are also quite general and quite necessary to the reproduction of the mode of production. The result – with apologies to Marx, Baudelaire, and Benjamin – is that "all that is solid melts into space;" but capitalism, transformed and transported to new places, remains.

Finally, this book may be read as social theory. Those who do not labor as geographers, urbanists, regional economists, or in any similar field, might, at this point, respond, "one surely does not need to read geography to learn about politics, economics or sociology." Yet geography is one of the best windows through which to view society at work, because it is an extraordinarily complex manifestation of societal relations and productive activity. As a unity of diverse determinations of social life, geography often requires the most sophisticated, agile theorizing. Geography inherently demands that theory be confronted with concrete outcomes, because important political, economic, and social processes are only determined in light of the ways they are embedded in place and spatial relations. Hence our form of spatial-temporal analysis should be of broad interest across the social sciences, and not just to those who specialize in cities, regions, or problems of industry.

What follows is inspired by the agenda and insights of Marxism. We are well aware of the lively debates currently swirling in and around Marxist social theory concerning the very fundamentals of method in social science. Without tackling the philosophical questions directly, we nonetheless try to walk the fine lines between various intellectual pitfalls. We recognize both the force of social structure and the initiatives of human agency; both necessity and contingency in the causation of particular events; both the widespread impact of deep structure and the relative autonomy of various parts of societies; both constancy and change in the capitalist mode of production; both the global and the local in the operation of economic systems; both the forces of production and class struggle as motors of history. Without pretending to have solved the deep puzzles of social explanation, we hope to limn a framework that is provocatively useful in advancing beyond mere restatement of dualisms to the creative application of a supple – dare one say dialectical? – social theory.

1 The Inconstant Geography of Capitalism

1.0 Geographical industrialization

Geographically uneven development and shifting relations of political and economic power between places are nothing new in human history. Dramatic changes in the arrangement of space economies and the creation and destruction of centers and peripheries have been features of civilization since ancient times. Greece, Rome, and Byzantium, for example, each established its own set of core regions, urban centers and economic peripheries. Each experienced considerable dynamism through military conquest, trade and settlement, as new cities were founded, flourished or fell into ruin (see e.g., Burns, 1966). Feudal Europe, less urbanized than ancient slave and tribute societies, established a relatively stable system of small towns, yet at the high tide of feudalism the geographical expansion of settlement, trade and conquest was remarkable, and the fortunes of places could vary quite sharply over time, as the rise and fall of Bruges, Aachen or Les Baux illustrate (Bloch, 1948; Vance, 1977).

The rise of merchant capitalism, extensive commodity production and the absolutist state during the period 1600–1750 brought a fundamental realignment of European urbanization and regional prosperity, from the Mediterranean south to the formerly backward Atlantic north (Braudel, 1972). This was accompanied by the greatest wave of new city foundings and urban growth since the eleventh century, including the spectacular and wholly unprecedented explosion in the size of the very largest cities such as London, Paris, and Amsterdam (DeVries, 1984) (see figure 1.1). Thus was the modern "urban revolution" launched with the entry of capitalism in its mercantile form on the European continent .

The development of industrial capitalism after 1750 set in train new long-term urbanization patterns marked by the spread of medium-sized factory towns, the overwhelming force of industrialization in city-building and regional growth, and, of course, the breakneck urbanization of Great Britain as the fast-growing heart of the industrial revolution (DeVries, 1984; Vance, 1977). By the middle of the nineteenth century, the spatial organization of capitalism had changed again, spreading rapidly to new areas such as northern France and eastern Belgium, the American Midwest,

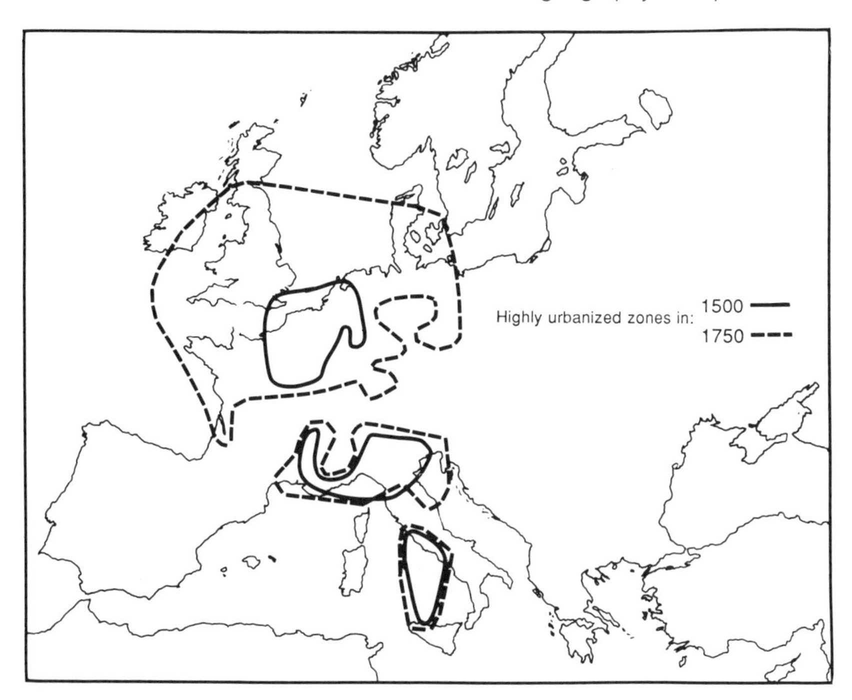

Figure 1.1 *Northward shift in European urbanization between 1500 and 1750* The Atlantic cities of the new commercial and merchant capitalist economy expanded dramatically while the Mediterranean world stagnated during this period. The boundaries indicate areas of roughly equivalent urban potential, or access to urban population centers. For precise explanation of the calculations, see DeVries (1984, pp. 154-67). (After DeVries, 1984)

and the German Ruhr, accompanied by the burgeoning of cities such as Paris, Chicago, and Milan (Pollard, 1981).

Flux in the space economy and city-systems of capitalism continues to this day, now driven primarily by the growth and locational dynamics of modern industries. To take just one example, the advent of the microelectronics industries has triggered a series of major, and largely unforeseen, shifts in the geography of North America. Electrical equipment and electronics production in the United States was originally located in the large metropolitan regions of the Northeastern seaboard, stretching from Massachusetts through New York and New Jersey, and as far south as Maryland. By the mid-1950s, however, the industry was expanding at a rapid pace both absolutely and relatively in the Southwestern states from California to Texas. The Northeast remained an important center of electronics production, but its locational primacy was decisively lost by the 1960s. The central sector of the microelectronics industry ensemble, semiconductors, developed its principal production center in California's Santa Clara Valley. With the further growth of the electronics ensemble by the 1970s and 1980s, high-technology production had become a specialty of the Southwest. In turn, the growth of the industry within the Sunbelt has been characterized by two important countervailing locational tendencies: on one hand, these high-technology industries have evinced a definite tendency to spatial agglomeration in a series of emerging new growth centers like Silicon Valley, Orange County, Dallas-Fort Worth (or, in the Northeast, Route 128). On the other hand, there has been considerable dispersal of secondary production processes, such as wafer fabrication, in a wide swath of states along with the "off-shoring" of assembly operations to selected Third World locations (Scott and Storper, 1987). From the late 1960s to the early 1980s, so many formerly thriving industrial regions fell into stagnation that it is possible to speak of a major reorientation of the economic geography of North America. Major new cities such as San Jose and Anaheim have been added to the nation's urban system, while some older centers have declined in importance or have experienced dramatic changes in their population mix. Inter-regional linkages and transportation systems generally have tilted toward the South and West. The particulars of this story reflect the historical circumstances of both the United States and of the newer industrial ensembles, but a characteristic process can be recognized that underlies the spatial development of all advanced industrial countries.

In contemporary capitalist societies, economic development is principally the outcome of productive activities organized in the form of industries. The development of these activities and their locational dynamics are responsible for urban and regional development. What is unique about spatial development under capitalism is that it is carried out by individuals and private firms employing wage and salary workers, acting under conditions of generalized market exchange and the spur of competition.[1] These social foundations of capitalist production encourage the technological dynamism of the system, which in turn generates the new products, processes and inputs characteristic of industrialization. The expansive qualities of industrialization create the locational capability of industries, freeing them at crucial moments from the limits of the past and its geographical imprint. But industrialization, as linked to capitalist competition and the dynamics of capital accumulation, also generates a highly disequilibrated form of growth that repeatedly unhinges the existing economic order. We shall argue that the developmental dynamic of capitalist industrialization, not the allocative functions of its markets, drives geographically uneven development in capitalist societies. This developmental dynamic is specific to capitalism; the particular processes and forms of geographically uneven development differ in other types of societies.

[1] Even large state-owned enterprises in capitalist countries ordinarily end up conforming to the contours of international competition and "efficient" resource allocation.

To comprehend the nature of this process, we need to follow capitalist industries as they develop and weave a geographical tapestry of growth and decline. We call this **geographical industrialization**. It has four principal moments: localization, clustering, dispersal, and shifts. First, industries are often highly localized for much of their history; they neither arise ubiquitously nor spread out evenly among cities and regions. The result is that each territorial economy has its characteristic specializations. Second, certain locales within industries normally outrace the others as more and more activities that are part of, or related to, their production system cluster together. The resulting territorial complex of industry may be densely packed within a large city or spread more widely over an extended region. Third, industries eventually disperse some of their production units away from territorial clusters, at a variety of spatial scales. Fourth, new or radically restructured industries, with distinctive product lines and production methods, usually take up new locations, often outside previously industrialized regions.

Looked at from the point of view of places, the growth process is one of **territorial development**. New places are woven into the design as industrialization expands the dimensions of the fabric, while previously industrialized places fade in comparison. Territories – whether cities, regions, or countries – develop through sequences of successful industrializations, which reverberate from one place to another. Every broad epoch of territorial growth is based on a dominant ensemble of production sectors, such as textiles in the early nineteenth century, steel and machinery in the late nineteenth and early twentieth centuries, or automobiles and consumer durables in the middle decades of the twentieth century. An ensemble is dominant when its component industries exhibit most of the following features: employ large numbers of workers; absorb large amounts of investment; have unusually high rates of growth of output and/or employment; have major propulsive effects on upstream sectors; produce capital goods with critical effects on the products and processes of other sectors or produce widely-used consumption goods.

The advent of new industries and industrial ensembles also ushers in new "regimes of capitalist accumulation," or epochs of economic and social development (Aglietta, 1976; Lipietz, 1987). New industries introduce dramatic changes in employment relations, occupational structure, and income shares; these new arrangements of production and distribution, in turn, alter the nature of inter-regional and international economic relations. Novel forms of urbanization, daily life, and political culture arise in association with the restructured economies of capitalism. Such innovations in economic and social life are first developed in specific places and only later spread throughout the capitalist world through the processes of geographical industrialization, accompanied by inter-place competition and imitation. The economic and social relations that are dominant in world capitalism at any particular time are, in fact, outcomes of its historical geography: the history of capitalism is simultaneously its geography.

The dynamics of geographical industrialization and territorial development generate three general characteristics of the historical geography of capitalism: expansion, instability, and differentiation. Capitalist societies tend to be spatially expansionist in the same sense that they are economically expansionist. They produce new industries with some regularity and these tend to invade new places, to form growth centers where none exist. As these new industries themselves reorganize and decentralize, they may in turn create new growth peripheries in relatively undeveloped areas. Spatial shifts of this nature are everywhere in the history of capitalist industrialization, and are amplified by subsequent dynamics of territorial development. Second, the economic status of cities and regions is unstable over time. The creation of growth centers and peripheries, and the obsolescence of older industrial places, leads cities to undergo differential growth, and alters the configurations of city systems. Moreover, as major new industrial ensembles appear from time to time, there are major macro-regional shifts in the focus of capitalist growth, whether within a particular country or between countries. Third, as industries localize, cluster, disperse, and shift, they do so in a manner that continually differentiates territorial economies, in terms of output, employment, income, rate of growth, and a host of concrete qualities of local economic, political, and social structures. As a result, the historical geographies of countries tend to be very different from each other, and to manifest various levels of internal geographical differentiation.

These three dimensions of capitalist historical geography – expansion, instability, differentiation – are the outcome of the processes of geographical and territorial industrialization analyzed in this book. Before proceeding, we focus in greater detail on each of the three, to show how prevailing theories of industrial location and urban development fail to cope adequately with the realities of capitalist spatial dynamics.

1.1 Expansion

Modern capitalism ushered in the systematic industrialization of the economies of an ever-expanding range of territories, marked by the establishment of factories, plantations, and other consolidated sites of work, the introduction of large-scale mechanized production processes, the installation of wage-work and labor markets, and the insertion of huge quantities of commodities into both world trade and burgeoning local markets. The rate and scale at which all this has happened over the last 200 years had dramatically increased (Wolf, 1982). The industrial revolution began largely in one country, and only selectively there.

It quickly leapt over the English Channel to ignite parts of Belgium, France, and Germany, then moved on to southern and eastern Europe, sweeping over Russia by the end of the nineteenth century (Pollard, 1981) (see figure 1.2). In North America a continent was rapidly conquered by the force of an expanding capitalist society, allied to a commodity farming sector (Post, 1982). Beginning in New England, industrialism was in full swing throughout the Northeast by the 1850s, then surged westward to the Ohio Valley and Great Lakes region during the second half of the nineteenth century, creating a northern manufacturing belt that extended from Minneapolis to Boston (DeGeer, 1927). Early twentieth-century expansion came chiefly in the Old South and Pacific Coast, then spread to the Southwest, Florida, and some Rocky Mountain areas in the post-war period (Jusenius and Ledebur, 1978). The progress of territorial extension can be seen in the development of US cities over time (see figure 1.3). Japan entered the fold of industrial nations after the Meiji Restoration, and soon imposed a ruthless form of developmental colonialism on other East Asian countries, such as Korea, Taiwan, and Manchurian China (Cumings, 1984). Today the burgeoning industrial cities of Latin America, such as São Paulo, Mexico City, and Buenos Aires, approach or surpass the great metropolises of the North in population.

The expansion of capitalist territory occurs in such dramatic fashion that it brings about startling shifts in global financial, mercantile, and industrial centers. As Braudel (1979) has indicated, for early modern Europe the center of capitalist accumulation shifted about 1500 from Venice to Antwerp, then briefly back to Genoa, and then decisively north to Amsterdam in the early seventeenth century and to London thereafter. The growth of the Midlands cities restored some prominence to English urban-industrial centers, but on the European continent the industrial core shifted east to the Ruhr complex of western Germany by the turn of the century, despite London's continuing financial preeminence (DeVries, 1984). Across the Atlantic, New York, which had taken the leading position in the US system of cities, rode this country's industrial revolution to the top of the international heap by the middle decades of the twentieth century, not only as a financial powerhouse but also as an industrial region second to none in the world. Today, while New York and London retain world significance, the greatest industrial centers – particularly Los Angeles, Osaka, and Tokyo – are on the Pacific Rim, with Tokyo poised to steal New York's mantle as the principal center of global capital accumulation. In other words, not only does the forward edge of industrialization push outward, enlarging the effective core territory of the capitalist world, but the central places of the world economy are usually associated with the newest of the principal countries or meta-regions which have most recently shifted into industrial high gear.

New industrial ensembles seem to "leapfrog" in space, establishing growth centers and new growth peripheries that are relatively insulated

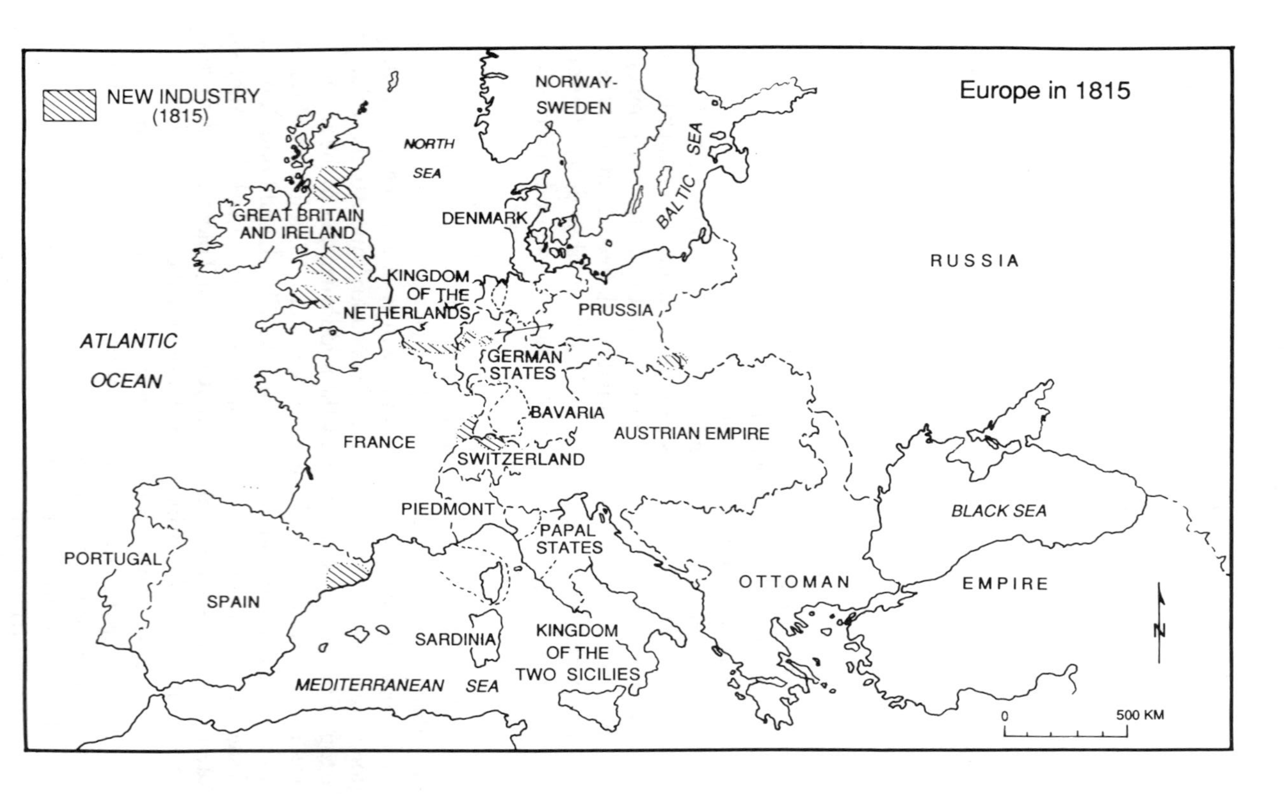

Europe in 1815
NEW INDUSTRY (1815)
NORWAY-SWEDEN
NORTH SEA
BALTIC SEA
GREAT BRITAIN AND IRELAND
DENMARK
KINGDOM OF THE NETHERLANDS
PRUSSIA
RUSSIA
ATLANTIC OCEAN
GERMAN STATES
BAVARIA
AUSTRIAN EMPIRE
FRANCE
SWITZERLAND
PIEDMONT
PAPAL STATES
BLACK SEA
PORTUGAL
SPAIN
OTTOMAN EMPIRE
SARDINIA
KINGDOM OF THE TWO SICILIES
MEDITERRANEAN SEA
N
0
500 KM

Figure 1.2(a)

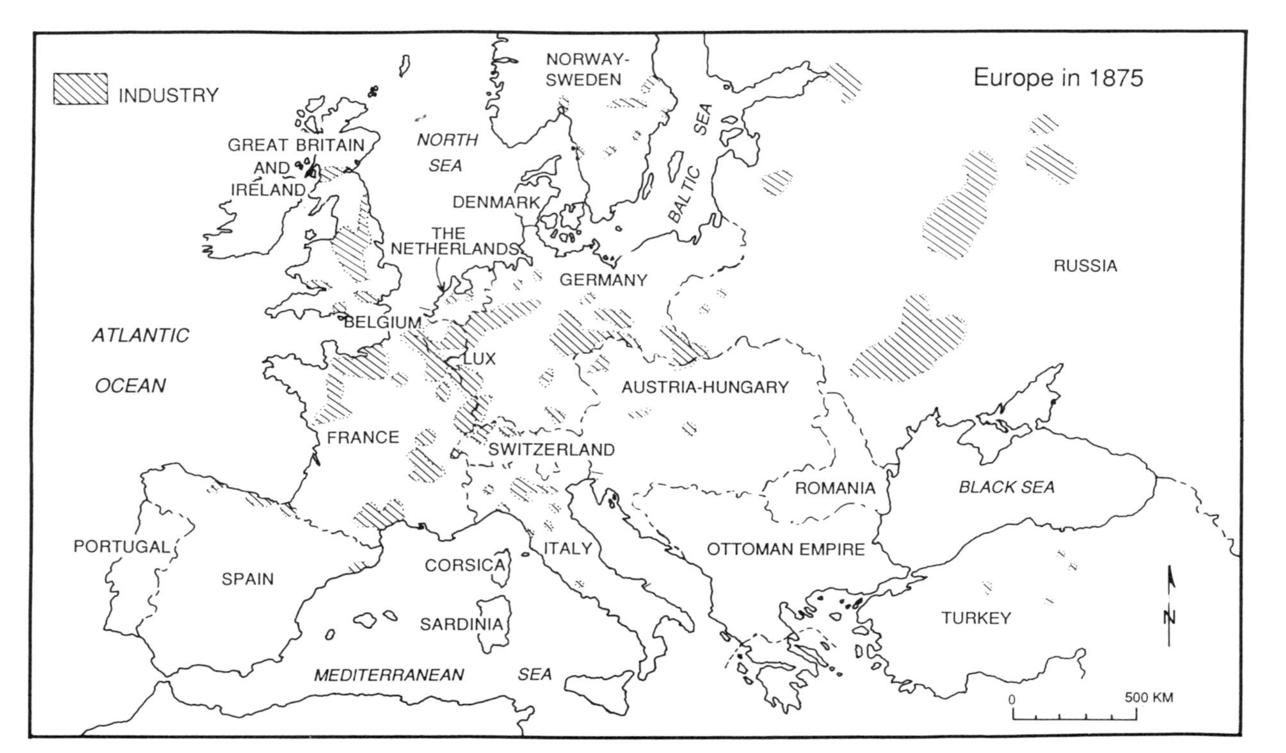

Figure 1.2(a,b) *Spread of industrialization in Europe, 1815–1875* The new Industrial Revolution, centered on textiles and iron-making, spread from its points of origin in Britain to a few places in Belgium, eastern France, western Germany, and Catalonia. From there it exploded across Europe, to such distant peripheries as Sweden, Russia and northern Spain. (After Pollard, 1981)

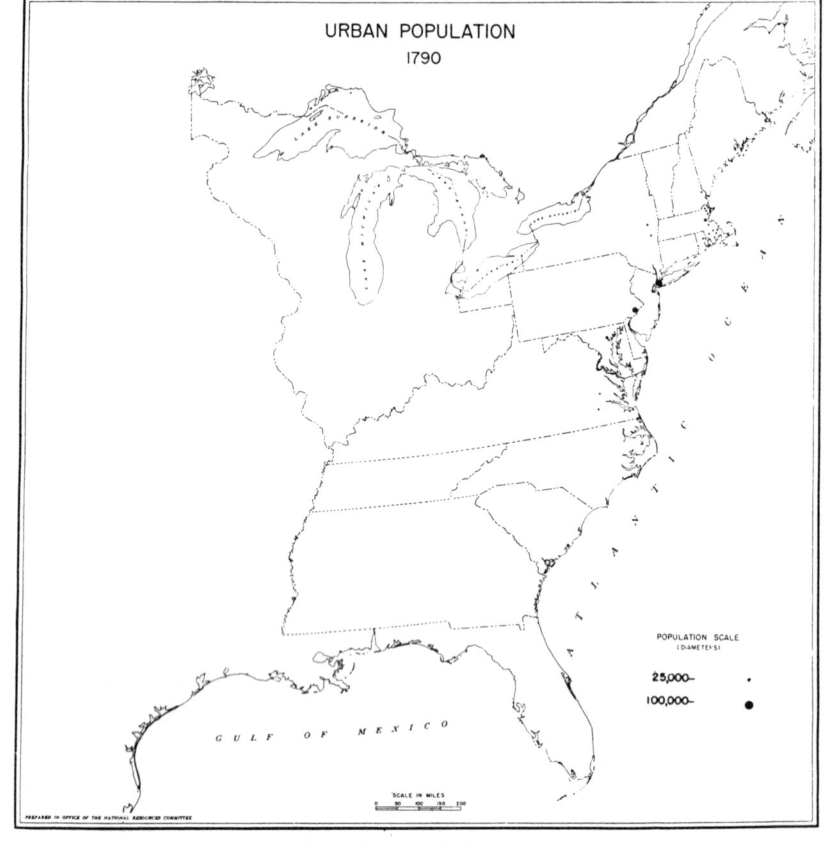

Figure 1.3(a)

from existing highly industrialized regions. This may involve the rise of new industrial nations or new regions within nations. In the latter case, rising industrial areas tend *not* to be in the largest centers of a nation's urban hierarchy, but are more typically in or near well-developed secondary urban-industrial regions (Taylor, 1915). Aerospace (near Los Angeles), electronics (near San Francisco and near Boston), electrical machinery (hinterlands of New York City), and steel (South Chicago and Gary) all developed as satellites or "suburbs" of major cities and helped expand their metropolitan dominance. All became the centers of important new regional growth processes, as the industries themselves developed through market widening and standardization of production (Alonso and Medrich, 1978).[2] Occasionally, they are located in cities even farther down the

[2] While much of the industrial location literature shows a strong correlation between urban size and industrialization, it rarely recognizes the causal force of industry growth in creating cities and city systems (for a critique see Scott, 1986, 1988a).

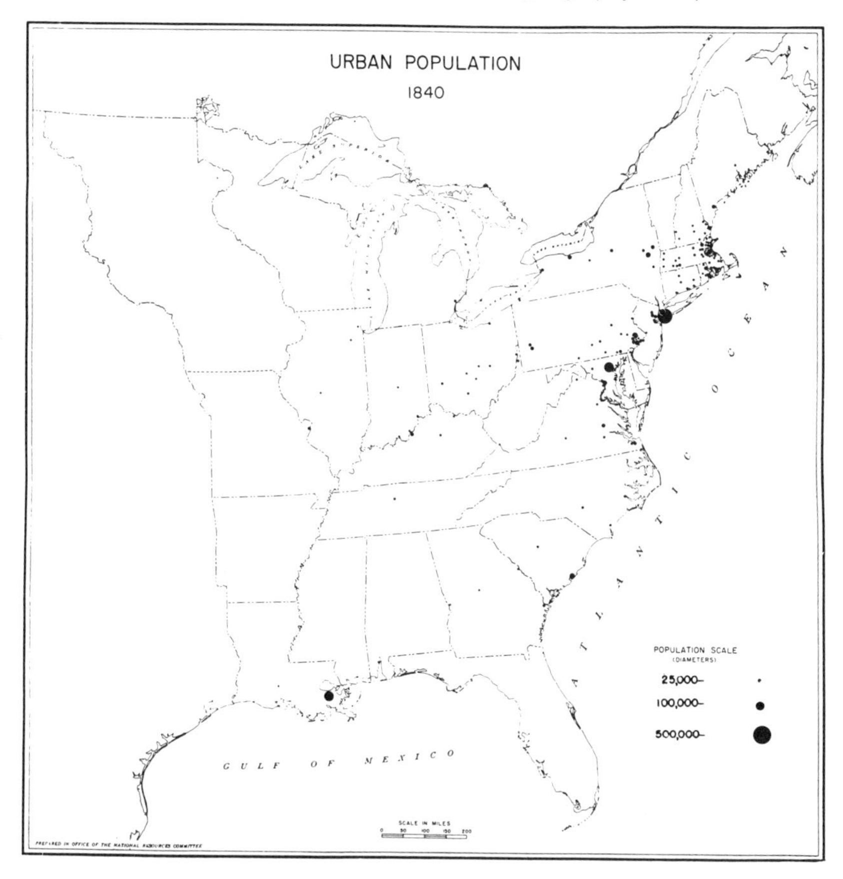

Figure 1.3(b)

national urban hierarchy, on a third tier, as with automobiles in Detroit, meat packing in Chicago, flour milling in Minneapolis, brewing in Milwaukee or wood products in Seattle (Boas, 1961).

The same general phenomenon can be seen at the international level, where the new powerhouses of development have generally *not* been among the leading national economies. This holds for the mercantile Netherlands (and Flanders) in the seventeenth century and Britain in the eighteenth, both formerly in the shadow of France, Spain, and the Italian peninsula; for Germany in the nineteenth century, which lagged behind the rest of Europe in national integration before Prussian conquest; for the United States, which was only an outpost of European civilization before its mid-nineteenth century rise to economic prominence. No one predicted the stunning rise of Japan from the ruins of World War II. Germany, the United States, and Japan were all objects of scorn for their

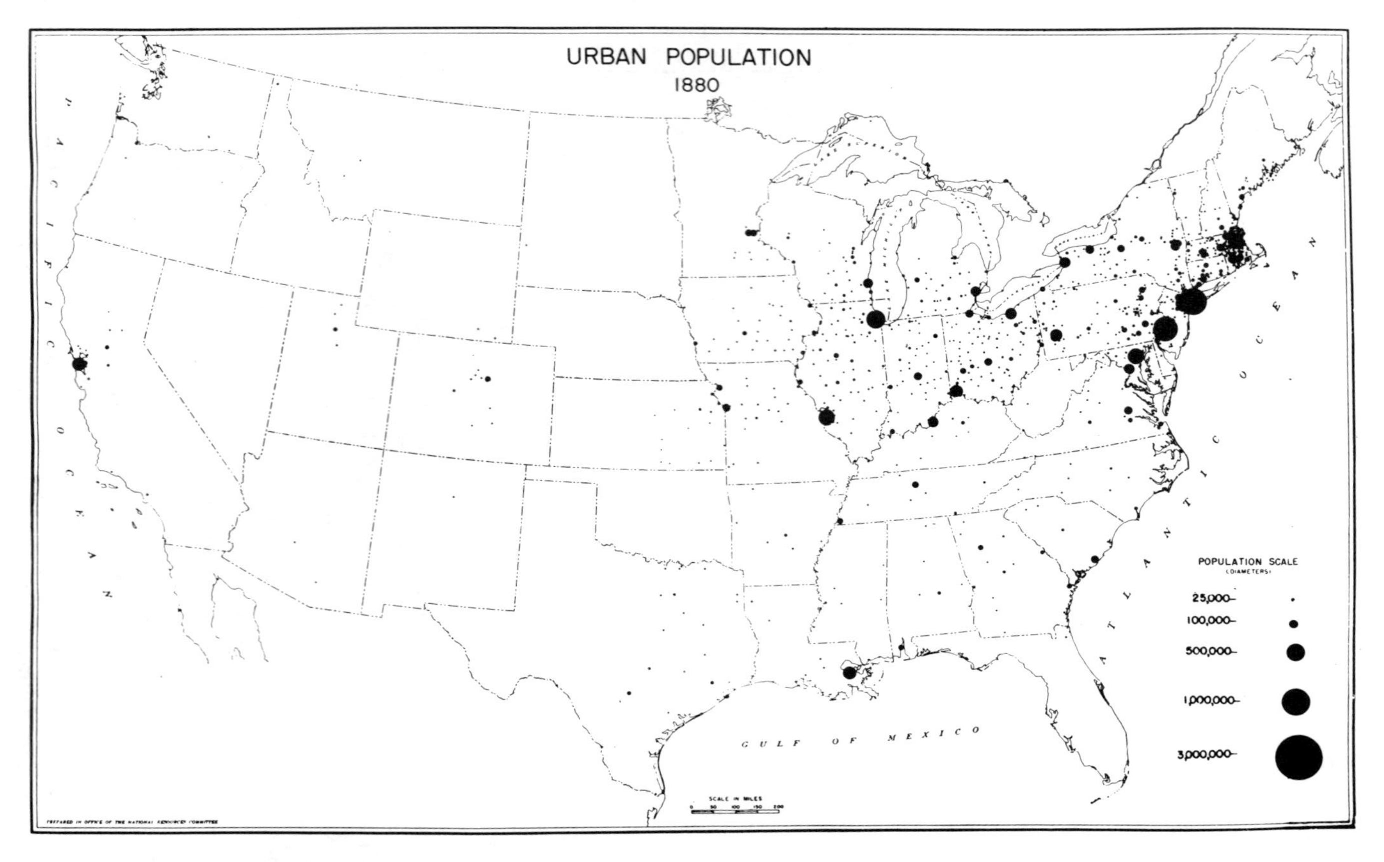
URBAN POPULATION
1880
PACIFIC OCEAN
ATLANTIC OCEAN
GULF OF MEXICO
POPULATION SCALE
(DIAMETERS)
25,000
100,000
500,000
1,000,000
3,000,000
SCALE IN MILES
0 50 100 150 200
PREPARED IN OFFICE OF THE NATIONAL RESOURCES COMMITTEE

Figure 1.3(c)

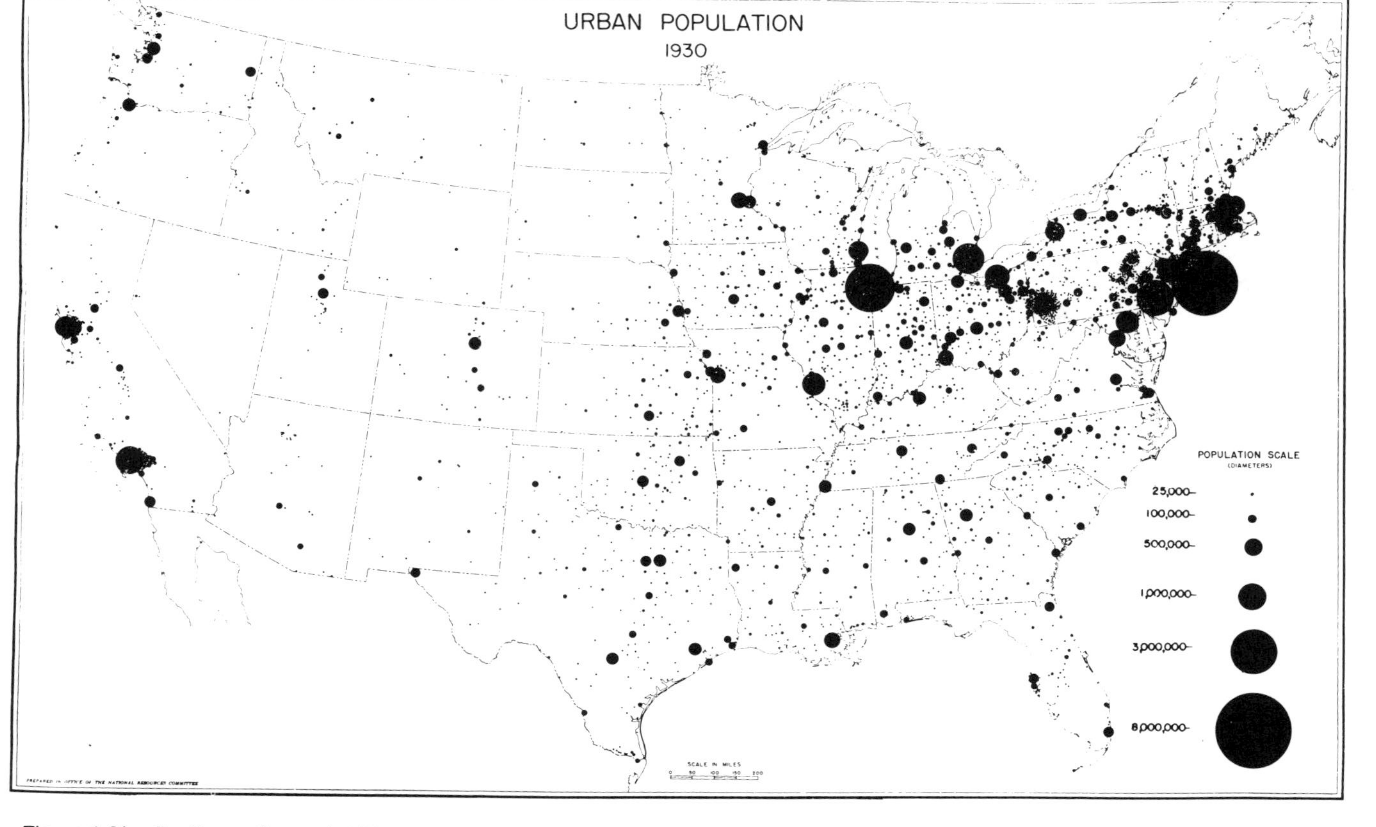

Figure 1.3(a–d) *Expanding and shifting pattern of urbanization in the United States, 1790–1930* Because the westward expansion of the US urban system is so clear, the shifting weight of cities and regions can be overlooked. Easily seen on these maps is the rapid rise of New Orleans and Baltimore before 1840, of San Francisco, Cincinnati or Buffalo by 1880, of Los Angeles and of the Midwest - especially Chicago and Detroit - by 1930. (See tables 1.2 and 1.3 for names, population, and rankings of US cities 1800–1985.) (From National Resources Committee, 1937)

shoddy goods and odd backward ways in their early stages of industrialization, but had the last laugh on their detractors.

The evidence suggests that there is something terribly wrong with existing theories of spatial development, none of which provides a ready explanation for rapid growth at the periphery. Neoclassical models focus on the stability of market economics and their ability to adjust to external events, such as technical change. For any given set of production technologies, income levels, and population, there is an equilibrium geography, i.e., a stable set of plant locations, urban centers and trade relationships (Isard, 1956, 1969). If firms are subject to perfectly competitive conditions, and household incomes and factor supplies are fully allocated, there is no endogenous source of change in the economy. Because resources cannot be siphoned off for speculative purposes in such an equilibrium world, change must come gradually, if at all. This is one reason why neoclassical economics has never proposed a satisfactory model of technological change. The same applies to the development of space: it is hard to imagine, from the neoclassical standpoint, that dramatic changes in spatial development could occur, since peripheries are likely to present firms with higher production costs than existing centers. As a result, neoclassical development models are chiefly of the "trickle down" variety, in which the leading territory gradually expands its boundaries, bringing the gift of growth to less favored regions (Borts and Stein, 1964).

Conventional efforts to introduce technical change, to provide a degree of dynamism to regional development models, are no more successful in coming to terms with extensive growth at the periphery. According to innovation diffusion theory, new techniques and industries incubate in well-developed areas and then move outward and down through the urban and regional hierarchy (Berry, 1972; Brown, 1980). Product cycle theories are similar in portraying young industries as arising in core regions and decentralizing as they mature, owing chiefly to a standardization of products and production methods that allows the penetration of distant markets and use of cheaper, unskilled labor in peripheral areas (Vernon, 1966; Hirsch, 1967). Neither of these models can account for outlying areas that outpace core industrial zones in rate of growth, however.

A partial way out of the dilemmas of trickle-down models is the export-led theory of regional growth put forward by North (1955). Returning to the classical principle of growth through division of labor, as embodied in international trade theory (Ohlin, 1939), North argued that peripheral regions could develop through specialization in certain favorable exports, as the Midwest did in grains. But this theory says nothing about the process of industrialization beyond the fact that some outputs are exchanged across long distances, leaving us to wonder how the Midwest went from grain exporter to the leading industrial territory of the early twentieth century.

Another approach to regional development which tries to escape the limitations of general equilibrium theory suggests that peripheral regions begin to attract industry rapidly once they pass scale thresholds necessary to support markets for particular activities (Lampard, 1955). This model makes population movements exogenous, however, begging the question of the economic basis of migration. All subsequent industrialization is left to multiplier effects, in a sort of "regional Keynesianism," resting on autonomous consumers, income flows, and market thresholds. Development is reduced to the location of branch plants attracted to outlying areas by growing markets; the real dynamics of industrialization are ignored. A variant of this theory sees the implantation of branch factories leading to a sufficient concentration of activity to act as a new incubator, followed by a sort of spontaneous combustion of industrial spinoffs (Norton and Rees, 1979). This is similar to Rostow's (1961) international theory of the take-off to industrialism which also has recourse to mysterious "thresholds" of economic activity and innovation in the center. But the evidence is that places such as the Midwest or California were not branch-plant economies before they entered their greatest periods of growth; that economies of scale occur all through the process of industrialization, not only after a certain level of maturity; and that innovators do not emerge predominantly in the old core areas.

The evidence also refutes the once-prominent dependency thesis that backward regions and nations can never break the vise of underdevelopment and domination (e.g., Frank, 1969). This is not to deny that innumerable obstacles face poor countries trying to develop, including colonialism and imperialist interventions, nor to argue that industrialization spreads evenly or rapidly in all directions, but simply to observe that many formerly backward regions have successfully industrialized (Foster-Carter, 1985).

In short, to explain dramatic expansions of industrial space effected by dominant production ensembles we must abandon notions of hierarchical diffusion, equilibrium analysis, take-off points or dependency. Territorial industrialization does not consist in a trickling down of growth impulses, smooth adjustments and expansion in equilibrium, the servicing of new markets based on footloose populations, or the permanent dominance of the developed over the underdeveloped. Industries are able to create a productive capacity that did not exist before, often without very much regard to the previous conditions of the place in which they are situated. To a large degree, they provide their own impulses toward development, endogenously, in place. That is fundamental to the nature of industrialization as a world-historical – and world-*geographical* – process.

1.2 Differentiation

Capitalist industrial countries – and territories – manifest a great variety of industry localization and urbanization patterns. Yet even though there

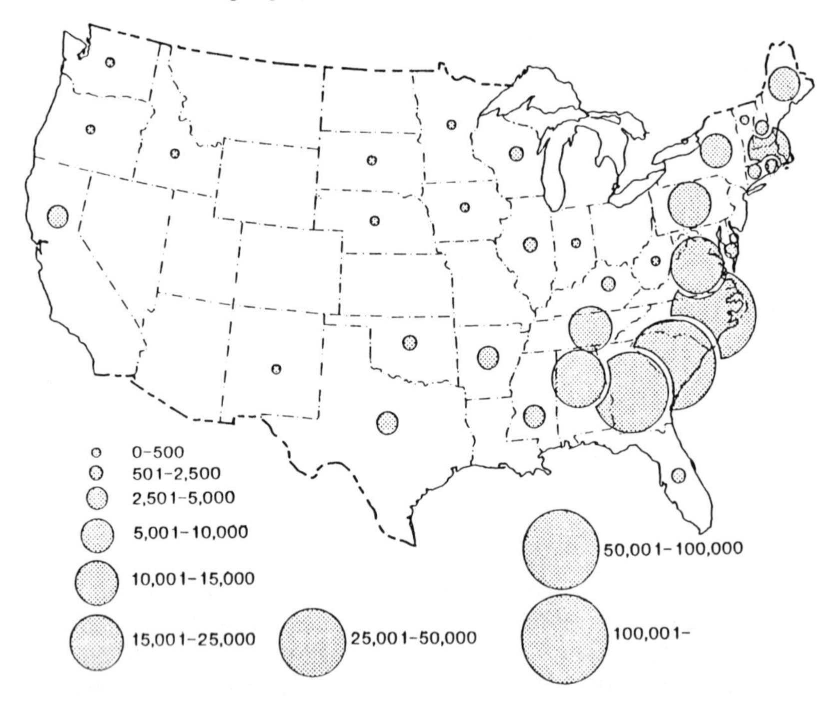

Figure 1.4 *Localization of the US textile industry, 1969* The regional concentration of textiles in the Southeast is readily apparent, despite the aggregation of several specialized (and more localized) product groups, from carpets to cotton goods. Circles show employment of wage-workers, by state, in SIC categories 2211, 2221, 2231, 2251, 2252, 2257, 2258, 2271, and 2279. (Data from US Bureau of the Census, *Census of Manufactures*, 1970)

is no common overall pattern of development in the capitalist countries, nor in industrialized countries generally, one common feature of development among them is the persistent tendency for regions to be economically differentiated from each other. That is, regions tend to be highly specialized in terms of their economic bases and, as a corollary, most industries show a marked skew of their activities towards relatively few places, especially within any one country – hence the common identification of Michigan with motor vehicles, North Carolina with textiles, Texas and Louisiana with oil and refining, or Wall Street with securities dealing (see figure 1.4). In fact, the degree of localization is often much finer than these aggregated definitions suggest: in US textiles, for instance, each subsector has a distinct center, whether it is carpets in Georgia, white goods in North Carolina or industrial fabrics in New England. Some industries, such as large-draft ship-building or earth-moving equipment, serve the entire world from a handful of places, indeed, a few have only one or two significant centers of production as with high-grade plywood in the Pacific Northwest of the United States and Canada or special-effects films from London and Los Angeles.

This localization of industry has traditionally been explained, in the classic theory of location, chiefly in terms of transportation costs for raw materials and finished goods (Weber, 1909; Hoover, 1948; Smith, 1981). By this explanation, improvements in infrastructure ought to have reduced the degree of localization of industry by lowering costs of transportation and communication, but there is no reliable evidence to support this. Instead of a new generation of "footloose" industrial plants, new industries have simply brought about new patterns of localization (see e.g., Cooke, 1986). This geographic specificity still requires explanation. It cannot be handled in the traditional terms of access to a given spatial distribution of pre-existing factors of production however. Instead, we shall have to look into the way industry produces its own conditions of localization.

Other patterns of uneven territorial development are widely recognized, above all the concentration of economic activity and population of capitalist nations in densely built-up cities. It is a commonplace to note that urbanization has always accompanied capitalist development, from eighteenth-century Sweden to twentieth-century Venezuela. The degree of urbanization (depending on the exact definition of urban places) approaches the truly remarkable figure of 95 percent in some of the most advanced industrialized regions, such as England and California. Yet some argue that urbanization is losing its grip on humanity, as new technologies and social practices make it easier to disperse the population (Webber, 1964). Industrial decentralization from large cities has been taken for granted for decades and the only dispute has been over the causes: the shift to truck and air transport (Chinitz, 1960), the economic costs of density (Berry, 1976), the prior suburbanization of the employed populace (Mills, 1972), lower wage rates at the periphery (Kain, 1968), or the space demands of single-story factories (Hamer, 1973). Such views follow from Weberian location theory which emphasizes the role of cost minimization in site selection. And behind this lies the predilection of neoclassical economics for spatial equilibrium, in which the long-term tendency is toward a dispersal of activity owing to the way cities drive up factor costs.

There is, however, almost no factual basis for claiming that the pull of urban areas on industry and people has diminished. Despite repeated predictions of diseconomies of scale, no upper bound has been hit to the size of big cities, even though metropolitan regions have spread out more than ever before. Tokyo, Mexico City, New York, and São Paulo all today contain close to twenty million souls. Three cities, Mexico City, São Paulo, and Buenos Aires, together generate over 50 percent of the manufactured output of all Latin America (Portes and Walton, 1981). Los Angeles has a gross domestic product greater than all but eleven countries (Soja, 1986). These huge urban agglomerations appear, moreover, in the newest and fastest-growing territories of the capitalist system, not just as products of continuing urbanization in older cities. We offer an explanation for the persistence of urbanization as the primary form of uneven geographical development based on the dynamics of industrialization, and argue that the force of agglomeration is as strong as ever, even in an age of instantaneous communication and well-developed transport.

Another long-standing puzzle of uneven development occurs at the level of very large territories: why do countries which are similar in terms of per capita income and other measures of economic output persistently manifest differing levels and patterns of urbanization? In the midst of the ever-increasing spatial integration of the capitalist world, space-economies continue to differ in important ways. Yet much of the literature on national urban development and modernization holds that countries with similar levels of economic development, income, and industrial activity will develop similar levels of urbanization (Bairoch, 1977; Kelley and Williamson, 1984). The literature explicitly concerned with space goes even further, claiming that countries should ultimately develop similar patterns of urbanization (El Shaks, 1972; Alonso, 1980). This thinking typically rests on the prevailing notions of neoclassical economics about tendencies toward spatial equilibrium, but it may also derive from simple schematic models of "modernization" or physical analogies such as "gravity models."

The most important empirical formulation for convergent urban systems is the rank-size rule, popularized by Zipf (1949). Building on the commonsense observation that there are fewer big cities than small ones, it defines the precise distribution of the number and size of cities for a given population (or income level).[3] The rank-size rule retains a strong grip on the imaginations of social scientists because it brings simplicity and order to the very complex phenomenon of urbanization and regional development (Gregory, 1978), although, paradoxically, research attempting to prove rank-size ordering is a rich source of evidence as to the actual diversity of spatial patterns from country to country. Table 1.1 provides hypothetical cases of different city systems generated by allowing the exponent *q* to vary (see footnote 3). An exponent less than one generates a relatively large number of big or middle sized cities while an exponent greater than one creates an exaggerated "primate city" at the top of the hierarchy. The middle column describes the classical, or restrictive,

[3] Starting with the largest city, a rank-size order can be found for each succeeding level of cities, grouped by size and share of the total population. The rank-size rule simply applies a generic statistical distribution to urban populations. This alone should make us suspicious about its veracity, unless we are prepared to stipulate that population movements are primarily the outcome of stochastic processes, as does Berry (1961). The classic rule states that differences in the population of cities of different sizes is equal to a constant ($p_i R_i = K$) where p_i is the population of the city class "i," R_i is the rank of class i ordered by population size, and K is a constant. This restrictive rank-size rule can be relaxed by introducing an exponential function of the form $R_i p_i^q = K$, or converting it to a logarithmic function (lognormal distribution) of the following form: $\log R = a' + b' \log p + c' (\log p)^2$.

Table 1.1 Hypothetical city hierarchies

$p(R)$	$q = 0.20$	$q = 1.00$	$q = 1.80$
p^1	5,000,000	5,000,000	5,000,000
p^2	4,352,753	2,500,000	1,435,873
p^3	4,013,708	1,666,666	692,072
p^4	3,798,291	1,250,000	412,396
p^5	3,623,898	1,000,000	275,946
p^1/p^2	1.149	2.000	3.482
p^1/p^3	1.246	3.000	7.225
p^1/p^2+p^3	0.598	1.200	2.350
$p^1/p^2+p^3+p^4+p^5$	0.317	0.779	1.775

In these three possible rank-size distributions, the "primacy" of the largest city (p^1) increases in comparison to the size of the next four largest cities as the exponent, q, of the logarithmic equation (shown in fn. 3, p. 22) increases

case of $q = 1$. The restrictive linear version of the rank size rule (i.e., where the exponent is equal to one, as in the middle column of table 1.1) simply does not hold up. Rosen and Retnick (1980), for example, find the exponent values range from 0.809 for Morocco to 1.962 in Australia. In other words, Morocco has too many medium sized cities, Australia too many large ones.[4]

Thus, the shapes of national urban hierarchies differ. Some are dominated by one large metropolis with few intermediate sized cities; others have a big metropolis and a substantial system of intermediate sized cities; still others show no particular metropolitan dominance. Only in a very few cases is the restrictive rank-size relationship close to being borne out, particularly in Canada and the United States.[5]

[4] Another approach measures the primacy of the largest city in the system with respect to cities of intermediate size, for instance by calculating the size of the largest city as a percentage of the next two, three, or five largest cities (Zipf, 1949). For example (using data from the 1970s), Budapest accounts for fully 76 percent of the population of the next five cities in Hungary, London 70 percent of the next five in Britain, Paris 55 percent for France, New York 45 percent for the United States, and Toronto 35 percent for Canada (Rosen and Retnick, 1980). In a more expansive comparison, Budapest represents 45 percent of the top 50 cities' population in Hungary, while New York City contains only 19 percent of the top 50 in the United States. Cities of intermediate size are thus less important in Hungary than in the United States. In 30 of the 44 countries surveyed by Rosen and Retnick the distributions indicated greater-than-proportionate attractions for the larger cities, while 14 countries showed a tendency for growth to be directed away from the largest cities after they reach a certain size. Among the latter are the United States, Canada, West Germany, and the USSR; among the former, the other Western European countries and the more industrialized nations of the Third World.

[5] The same variability has been found for the early modern period in Europe. As DeVries concludes, "The most obvious result of this survey is the absence of any common pattern of national urban evolution . . . every urban hierachy is adapted to a specific society, topography, and technology" (1984, p. 120).

Researchers have attempted to explain these differences in the shape of a nation's urban system in various ways: higher-income countries are said to develop more even distributions of population, federal state systems to promote more dispersed urban centers and larger area-to-population ratios to encourage a more even distribution of cities. But none of the explanations works (Langenbruch, 1981). This led Berry (1961, p. 587) to conclude that "there are no relationships between type of city size distribution and either relative economic development or the degree of urbanization of countries, although urbanization and economic development are highly associated." Sheppard has gone still further, concluding after exhaustive statistical tests, that "the variables postulated by various authors to date almost completely fail to explain empirical deviations from the rank-size relationship using international data" (1982, p. 139).[6]

Yet even in abandoning attempts to prove the spatial convergence of development, we should not assume that geographical differences defy systematic explanation. Rather, a different kind of theory is wanted. In fact, gross aggregates such as industrial output, income, or population are very poor ways to get at the formation of territorial economies. Territorial growth can only be understood by opening up the "black box" of production to reveal its geographical dimensions. Industrialization is not all of a piece. Aggregate data, especially of such simple measures as population size, tend to wash out the interesting differences among places and to miss the particular handles which might give us a grip on the forces and relations of production that drive territorial growth. By analyzing the dynamics of geographical industrialization, we can account theoretically for a broad range of processes responsible for territorially uneven development, while at the same time fully accommodating the different patterns this development shows in different places.

1.3 Instability

The positions of cities and regions within urban systems, nations, and the international economy – in terms of population, employment, income, and output – are unstable under capitalism. Everyone knows of Britain's relative decline in this century, or of Japan's thunderous entrance into the first

[6] Lack of convergence can be observed with respect to industrial activity as well. For example, a much greater proportion of France's manufacturing output is concentrated in the Île-de-France region and the surrounding *départements* than in the New York metropolitan region for the United States (Verlaque, 1984). Patterns of change also seem to be imprinted with important national differences. For example, the primacy of the Île-de-France region has not substantially diminished despite significant government subsidies designed to promote industrial decentralization since 1954, while in the United States there has been a substantial redistribution of the nation's industrial base.

Table 1.2 The 20 largest cities in the United States, 1800 and 1840

	1800		1840	
1.	Philadelphia	61,559	New York	348,943
2.	New York	60,515	Philadelphia	220,423
3.	Baltimore	26,514	Boston	118,857
4.	Boston	24,937	Baltimore	102,313
5.	Charleston	18,824	New Orleans	102,193
6.	Salem	9,457	Cincinnati	46,338
7.	Providence	7,614	Albany	33,721
8.	Norfolk	6,926	Charleston	29,261
9.	Newport	6,739	Washington	23,364
10.	Newburyport	5,946	Providence	23,171
11.	Richmond	5,737	Louisville	21,210
12.	Nantucket	5,617	Pittsburgh	21,115
13.	Portsmouth	5,339	Lowell	20,796
14.	Gloucester	5,313	Richmond	20,153
15.	Albany	5,289	Lowell	20,096
16.	Schenectady	5,289	Troy	19,339
17.	New London	5,150	Buffalo	18,213
18.	Marblehead	5,211	Newark	17,290
19.	Savannah	5,146	St Louis	16,469
20.	Alexandria	4,971	Portland	15,218

The US city-system of this period was dominated by port towns, and by the big mercantile cities of the Northeast. But the city hierarchy was not stable. By 1840, New York had moved to the first rank and New Orleans had joined the top five; several new industrial towns such as Providence, Troy and Lowell, had appeared, along with a number of western cities, such as Louisville and Cincinnati; and a few older northern and southern places, such as Charleston and Salem, had disappeared fromthe list. (Population data from Taylor, 1967)

rank of national economies. But such movements occur at sub-national levels as well. During the past two decades, the urban system of the United States has been reshuffled rather dramatically: Pittsburgh, Cleveland, Baltimore, and St Louis, for example, have fallen down the list of the largest cities to be replaced by the likes of Houston, Miami, and San Diego; Detroit and Philadelphia recently dropped in rank as the San Francisco region became the fourth largest metropolitan area in the country; Chicago has given way to Los Angeles as the "second city." Historically the same upheavals are readily observable. A comparison of the top cities and towns of the United States between 1800 and 1840 demonstrates similar volatility. Previously vigorous port towns such as New Haven (Conn.) and New Bedford (Mass.) or Dumphries (Virginia), which made their names on whaling or tobacco, fell from grace as the newer inland ports, textile or iron towns, such as Cincinnati, Albany, Lowell, and Troy came into prominence (Taylor, 1967). Nineteenth-century Detroit, Los Angeles, Dallas, and Seattle were minor stars in the urban firmament which grew explosively in the early twentieth century (see tables 1.2, 1.3 and figure 1.5).

Table 1.3 The 30 largest cities in the United States, 1890-1985

	1890 (City + Major Satellites)		
1.	New York	3,129,000	(2,507,400)
2.	Chicago	1,167,400	(1,099,880)
3.	Philadelphia	1,244,400	(1,047,000)
4.	Boston	1,042,200	(448,477)
5.	St Louis	467,000	(451,800)
6.	Baltimore	434,400	
7.	San Francisco-Oakland	387,000	(299,000)
8.	Pittsburgh	364,600	
9.	Cincinnati	351,800	(296,900)
10.	Cleveland	320,000	(261,300)
11.	Minneapolis-St Paul	297,800	
12.	Buffalo	261,000	(255,700)
13.	Washington	244,600	(239,400)
14.	New Orleans	242,000	
15.	Detroit	212,100	(205,900)
16.	Akron	208,400	
17.	Milwaukee	204,500	
18.	Providence	188,600	(132,100)
19.	Newark	181,800	
20.	Albany-Troy-Schnectedy	175,700	(94,000)
21.	Kansas City	171,000	
22.	Jersey City	163,000	
23.	Louisville	161,000	
24.	Omaha	140,500	
25.	Rochester	133,900	
26.	Denver	106,700	
27.	Indianapolis	105,400	
28.	Syracuse	88,100	
29.	Columbus	88,100	
30.	Worcester	84,600	
	1920 (City + Major Satellites)		
1.	New York	7,089,900	(5,620,000)
2.	Chicago	3,039,900	(2,700,700)
3.	Philadelphia	2,227,600	(1,823,800)
4.	Boston	1,884,900	(748,100)
5.	Cleveland	1,099,400	(796,800)
6.	Detroit	1,030,400	(993,700)
7.	San Francisco-Oakland	864,200	(506,700)
8.	St Louis	839,700	(772,900)
9.	Los Angeles	742,000	(576,700)
10.	Baltimore	733,800	
11.	Pittsburgh	635,200	(588,300)
12.	Minneapolis-St Paul	615,300	
13.	Buffalo	557,500	(506,800)
14.	Cincinnati	497,900	(401,200)
15.	Milwaukee	457,100	

(*Continued*)

Table 1.3 (*Continued*)

16.	Washington	455,000	(437,600)
17.	Kansas City	425,500	
18.	Newark	414,500	
19.	Seattle-Tacoma	412,300	
20.	New Orleans	387,200	
21.	Providence	374,800	(188,600)
22.	Indianapolis	314,200	
23.	Jersey City	298,100	
24.	Rochester	295,700	
25.	Albany-Troy-Schnectedy	274,000	(113,300)
26.	Dallas-Ft Worth	265,500	
27.	Portland	258,300	
28.	Denver	256,500	
29.	Toledo	243,200	
30.	Columbus	237,000	
	1950 (Standard Metropolitan Areas)		
1.	New York	12,912,000	
2.	Chicago	5,495,400	
3.	Los Angeles	4,367,900	
4.	Philadelphia	3,671,000	
5.	Detroit	3,016,200	
6.	Boston	2,370,000	
7.	San Francisco-Oakland	2,240,800	
8.	Pittsburgh	2,213,200	
9.	St Louis	1,681,300	
10.	Cleveland	1,465,500	
11.	Washington	1,464,100	
12.	Baltimore	1,337,400	
13.	Minneapolis-St Paul	1,116,500	
14.	Buffalo	1,089,200	
15.	Dallas-Ft Worth	976,100	
16.	Cincinnati	904,400	
17.	Milwaukee	871,000	
18.	Kansas City	814,400	
19.	Houston-Galveston	806,700	
20.	Providence	737,200	
21.	Seattle-Tacoma	733,000	
22.	Portland	704,900	
23.	New Orleans	685,400	
24.	Atlanta	671,800	
25.	Louisville	576,900	
26.	Denver	563,800	
27.	Birmingham	558,900	
28.	San Diego	556,800	
29.	Indianapolis	551,800	
30.	Youngstown	528,500	

(*Continued*)

Table 1.3 *(Continued)*

	1985 (Metropolitan Statistical Areas)	
1.	New York	17,918,000
2.	Los Angeles	12,759,000
3.	Chicago	8,080,000
4.	San Francisco	5,803,000
5.	Philadelphia	5,787,000
6.	Detroit	4,592,000
7.	Boston	4,052,000
8.	Houston	3,606,000
9.	Dallas-Ft Worth	3,526,000
10.	Washington	3,494,000
11.	Miami	2,865,000
12.	Cleveland	2,773,000
13.	Atlanta	2,469,000
14.	St Louis	2,422,000
15.	Pittsburgh	2,334,000
16.	Minneapolis	2,262,000
17.	Baltimore	2,252,000
18.	Seattle	2,251,000
19.	San Diego	2,133,000
20.	Tampa	1,871,000
21.	Denver	1,828,000
22.	Phoenix	1,817,000
23.	Cincinnati	1,681,000
24.	Milwaukee	1,551,000
25.	Kansas City	1,499,000
26.	Portland	1,350,000
27.	New Orleans	1,333,000
28.	Columbus	1,287,000
29.	Norfolk	1,280,000
30.	Sacramento	1,256,000

Despite the continued high ranking of many of the most populous cities for over a century, positions have been reshuffled many times, and new cities have entered the lists with regularity. Note the appearance of Pittsburgh near the top and Syracuse near the bottom by 1890, of Detroit and Toledo by 1920, of Los Angeles and Youngstown by 1950, of Houston and San Diego by 1985. Note, too, the brief rise and fall of Jersey City and Newark, or the slow decline of Baltimore and Milwaukee. We have tried to correct for changing census definitions of urban areas by including major satellites of the biggest cities and counting twin or triple cities as one, as is now the practice. Figures for core cities are shown in parentheses, where relevant. Census counts are usually too low for the largest metropolitan areas (Pred, 1966, p. 13n). (Population data assembled from US Bureau of the Census, *Statistical Abstract of the United States*, various years)

Indeed, several early manufacturing centers in Britain suffered the fate of deindustrialization long before anyone had coined the term: once-bustling areas of North Wales, the Severn Valley, and Derbyshire are now quiet rural backwaters (Pollard, 1981). DeVries, who surveys the ups and downs of European cities in the early modern period observes: "in each period the cities that pushed Europe's urbanization forward were

Figure 1.5 *Locations of the 30 largest metropolitan areas in the United States, 1985* (See table 1.3 for 1985 population rankings)

distinct in number, rate of growth and location within the matrix . . ." DeVries (1984, p. 136). Very generally, there were dramatic reversals around 1650, 1800, and 1850 in favor first of large cities, then small ones, and then larger ones again (see figure 1.6).

The evidence once again does not accord with prevailing notions of urban hierarchy, which proclaim that cities enjoy rank-stability within an urban system. By this view, as new cities and peripheral areas are added, the urban hierarchy changes only in the sense that the base of the pyramid is broadened. The cities in the upper reaches of the hierarchy remain there; growth does not occur at the expense of older cities.

The claims of city hierarchy theory are important for two reasons. First, if the positions of cities really did tend to be stable, this would imply that, to a large extent, existing localities have nothing to worry about as capitalism develops, for they would virtually be assured of capturing a certain increment of that development. Second, at a more theoretical level, rank stability would imply that capitalist development is, to a large degree, imprisoned by its existing spaces and therefore that spatial-territorial change is not central to the dynamics of capitalism, as we have claimed. We have already suggested that empirically rank stability is not observable. But there are nonetheless powerful theoretical reasons for expecting development to be attracted to large existing centers of development.

There are three main explanations for city hierarchies, all premised on rank-stability. The "central place" model assumes perfectly hierarchical market areas, based on specialization in different orders of goods and

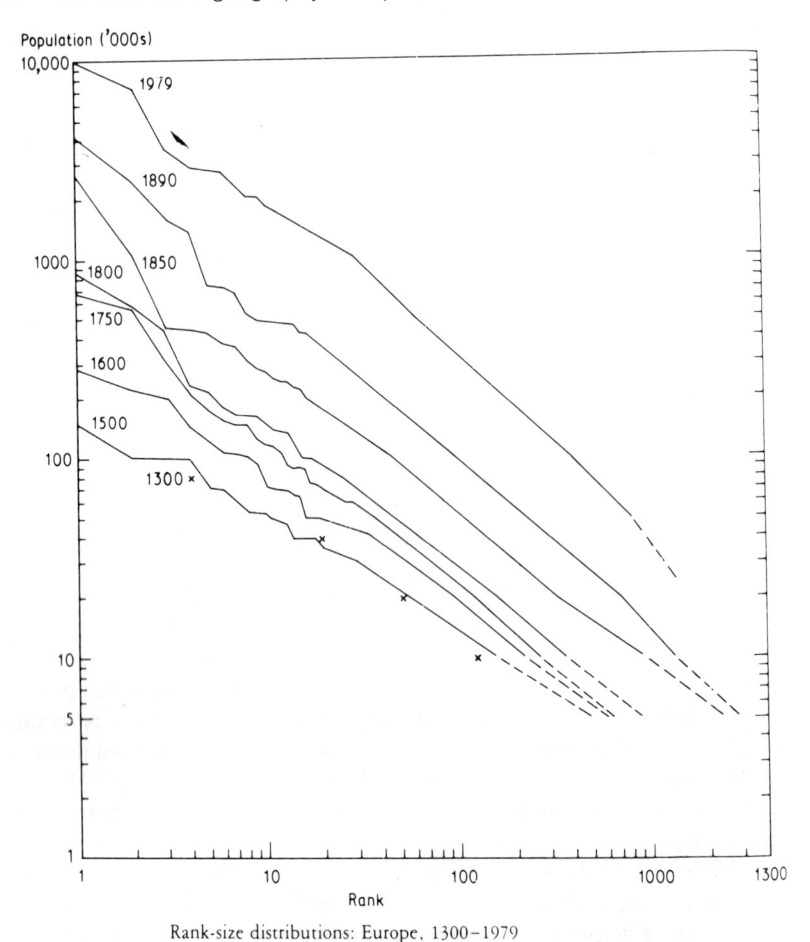

Figure 1.6 *The changing urban rank-size distribution of Europe, 1300-1979* The overall growth of cities after 1500 is readily apparent from the upward shift in the lines. Note, however, the accelerated expansion of the very largest cities between 1600 and 1750, indicated by the steepening of the gradient. In the following period of the Industrial Revolution, from 1750 to as late as 1900, there is a striking change as small and medium size cities outgrow larger ones - indicated by the flattening gradient of the lines (the widening gaps on the right-hand side of the graph). The big gap on the left edge of the graph between 1800 and 1850 signals the ascension of London to undisputed primacy among European cities. (From DeVries, 1984)

services (Christaller, 1935; Losch, 1944; Berry, 1967). Small towns provide basic items for the immediately surrounding populace, big cities capture those functions with large market areas (high minimum size thresholds for efficient provision of a good or service). The model also posits equilibrium, so by definition the largest city in the system remains the largest. Aggregate growth is a function of adding market area to the system. In principle, this growth could allow some older cities to expand by becoming more specialized, thus attracting activities formerly confined to a smaller number of large cities higher up in the pyramid, but the oldest and largest cities will become still more specialized, thus maintaining their positions in the hierarchy.[7]

A second model holds that city systems reflect not market territories but a new spatial division of labor based on hierarchies of labor, skill, and the command structure of large corporations (Hymer, 1972; Cohen, 1981; Massey, 1984). Headquarters and "higher" functions gravitate to the largest cities; divisional offices, mid-level engineering, advertising, and other services to secondary cities; and basic manufacturing ("branch plants") to tertiary sites. This portrayal of capitalist development stresses progressive concentration of production, capital, and political power in the largest multilocational firms, but juxtaposes this to a quiescent spatial order which the burgeoning corporations and their expanding internal division of labor simply reaffirm over time. There is, furthermore, no attempt for the most part to account for the fact that the urban hierarchy antedates the coming of the modern corporation.

The third and most interesting theory of city system development is rooted in the dynamics of spatial production. Pred (1966, 1974) offers a subtle blend of theory and history that sets a standard to which alternatives must be addressed. The Pred model is a particular version of the theory of circular and cumulative causation in economics, originally formulated by Myrdal (1957), which attributes regional disparities to aggregate factor productivity and income differentials. External economies in some regions open up productivity gaps between regions, which are then reinforced through differential increases in demand, the latter a consequence of the higher income that flows from higher productivity in the first place. These basic tenets are applied to the urban system, along with certain aspects of other theories previously mentioned. In this model, places that ultimately grow to become major cities gain an "initial advantage" by passing a certain size threshold, and come to enjoy perpetual advantage in capturing additional economic growth. Initial advantage is usually gained by successful specialization, as in export-base models, and Keynesian multiplier effects follow from the circulation of income derived from these export activities.[8] The locality goes on to assume additional central place functions, as growth encourages investment in infrastructure, which reinforces the transportation time and cost advantages of a budding

[7] Central place and Weberian location theory have been fully integrated, by Isard (1956), within the neoclassical framework in which every industry has a production function that allows for continuous substitution between input factors, including transport costs.

[8] Previous urban export-base models had been formulated as a simple mathematical relation between local multipliers and basic industries, with little to say about the development of city systems or long-run interurban relationships (Alexander, 1954).

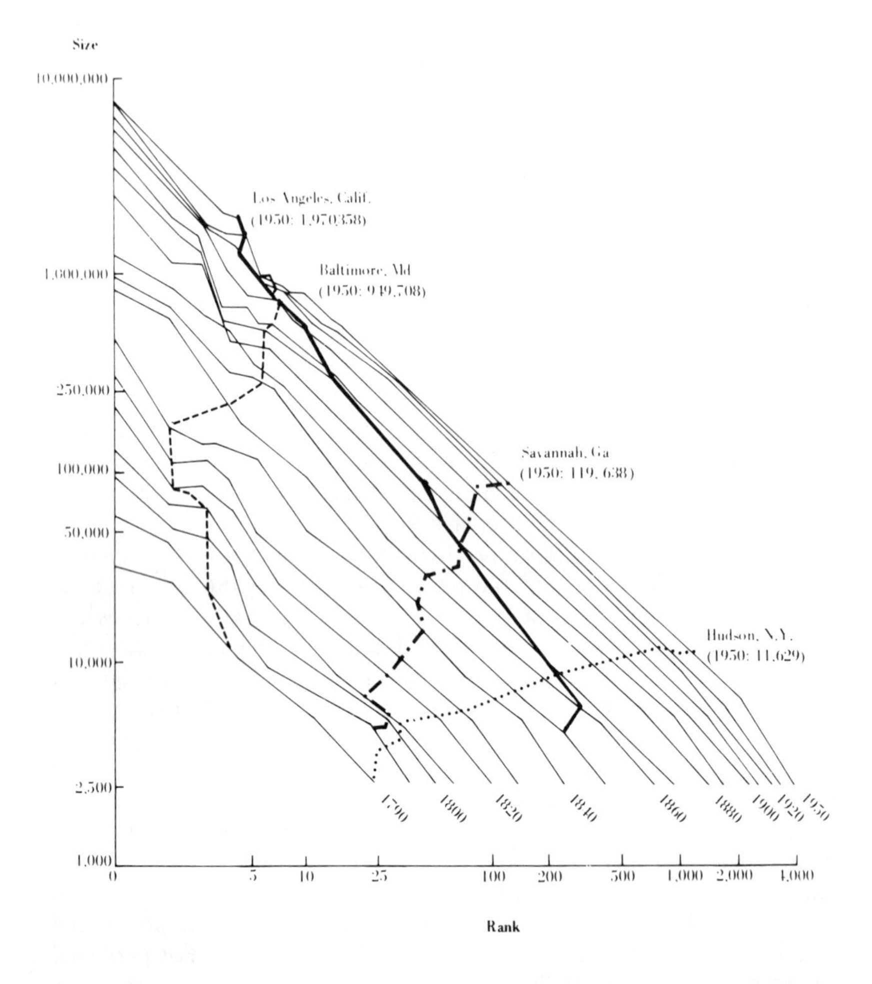

Figure 1.7 *Instability in the urban rank-size distribution of the United States, 1790-1950* Shifts in the overall rank-size order of US cities have been less marked than those in Europe. Nonetheless, the biggest cities generally grew faster than the rest in the early nineteenth century, followed by a gradual leveling up from below after 1830. Note the wide gaps, indicating rapid growth and rank instability, in quite different groups of cities in the periods 1830-40, 1840-50 and 1850-60. Much more dramatic are the shifting positions of individual cities, shown here for Los Angeles, Baltimore, Savannah and Hudson (After Dicken and Lloyd, 1977)

urban center. In turn, such cities become nodes of specialized information, which is presumed to be the source of technological innovations and thus future advantage. Concentration of higher order corporate functions adds further to the aggregate growth of the top-ranked cities (Holland, 1976; Pred, 1977). The outcome is circular and cumulative growth, with long-term rank stability for the system's largest units through these feedback loops.

We agree that major new units of the American urban system have arisen suddenly and grown rapidly in connection with their exportable economic activities (Wheeler and Brown, 1985), but differ on the impact this has on city rankings. The Pred model fails to anticipate sufficiently the rise of new cities up the urban hierarchy and the diversion of growth from cities that were supposed to enjoy permanent advantages. What Pred interprets as long-term rank stability, based on observing US urban development from the mid-nineteenth to mid-twentieth century is, in fact, medium-term stability associated with the locational behavior of two dominant groupings of economic activities mercantilism/textiles and steel/machinery, especially in the Northeast and Midwest. Moreover, the United States manifests an unusually stable hierarchy in comparison to Europe (DeVries, 1984). A slightly longer-term view than Pred's leads to rather different conclusions. Los Angeles is a case in point. Pred (1966) lists this as one of those cities with "initial advantage," but Los Angeles in fact came on the scene very late: it was no more than a cow-town before 1880 and had only 100,000 people by 1900 – and then outgrew every other metropolis decade after decade in an unbroken string for the next eighty years (see figure 1.7).

In this book we offer an alternative framework for explaining rank-size instability and other manifestations of upheaval in the space-economy of capitalism, in which new economic activities can create dynamic growth centers well outside of established metropolises. The location of these economic activities is not subject to the constraints described by the Pred and central place models. Market size thresholds matter little to an exporting industry. Cumulative causation can propel a new growth center more rapidly than an old one, if underlying industrial conditions are right. Transportation links can be built up as an area grows, as can labor skills, supplier firms, and other inputs whose costs appear to be prohibitive at the periphery. Nor is there any rigorous evidence, as Pred (1966) himself points out, that technological innovations are concentrated in the largest cities, necessarily giving them continuous leadership in the newest industries. Furthermore, the spatial biases in the transmission of information, so important to Pred's model, reflect the concentration of specialized, information-dependent activities in certain large cities; that is, information is specifically attached to activities such as mercantile trade, banking or integrated circuit design, not generally to places. Finally, whole new groups of cities and regions move to the forefront, in concert with broad epochs of capitalist growth based on new ensembles of dominant production sectors.

Cities that hold their ranking over the long term, on the other hand, do so by continuing to capture propulsive industries. In New York there has actually been a renewal of propulsive forces, as in the securities-trading and international financial sectors. Similarly, electronics and finance have kept greater Boston high up the metropolitan scale, despite a long period of industrial decline in New England. And even as these propulsive forces help maintain the region's overall standing, aggregate figures mask the rearrangement of both the industrial base and laborforce. In New York, Wall Street has elbowed aside the old blue-collar industries, such as garments, for supremacy in the regional economy (Gold, 1987). In New England one finds only the shadow of the mill economy in today's computer companies, even where they are located in the mill towns and in the mill buildings themselves (Harrison, 1984).

1.4 The need for an alternative theoretical framework

None of the prevailing models of industry location and regional development can cope with the expansion, instability, and differentiation that characterize the inconstant geography of capitalist industrialization. Massive deindustrialization of older economies such as Britain and the Northeastern United States should not have happened in an equilibrium world, nor in one driven by cumulative causation. One may ask corporate hierarchy theorists how Cleveland could lose half its population in the last decade while remaining among the country's top ten headquarters cities. In the 1980s, neoclassicists seized upon the selective booms in older areas such as Boston, North-Central Italy, and Jutland as evidence of lagged equilibrium adjustment processes but they cannot explain why all places with similar patterns of deindustrialization are not experiencing the same growth. Cumulative causation and dependency theorists have been surprised by the industrialization of countries such as Korea and Brazil, which were supposed to be locked into a low level of industrialization. And if spatial filtering were the essence of territorial development, how could empty California have risen up to exceed the national income of Great Britain in little over a century? There is nothing in the hierarchical conception of things to explain turning points in the fates of places, to show why, in some cases, the poor get richer and rich get poorer. On the other hand, optimistic neoclassical economists have been disappointed by the fitful nature of industrialization in the Third World, and the strategy of export-led growth has ignited full-scale industrialization in very few places. Nor has the industrial development of all peripheral areas in developed countries, such as the southern United States led to the convergence predicted by equilibrium theory. International trade models banish history altogether; barring an exogenous push, there is no reason to expect movement away from an existing equilibrium toward another condition with different spatial contours.

It now remains for us to demonstrate the causal efficacy of the processes that constitute geographical industrialization and territorial development in accounting for the inconstant geography of capitalism. The next five chapters focus on the anatomy of industry development and its spatial dimensions; the last two chapters come at the matter from the angle of the territorial basis for broader inter-industry and inter-class relations. We shall argue both for the generality of the forces behind many specific outcomes and the specificity of very general processes, especially their locational specifics, not only in the sense of a spatial imprint on particular places but also in terms of the imbrication of place and spatial relations in the very fabric of the industrialization process under capitalism.

2 Industrialization as Disequilibrium Growth

2.0 The production of growth

The analysis of urbanization, regional development, and industry location must be based upon growth theory. That is one fundamental reason for speaking of geographical industrialization rather than "industry location." The sites of economic activity are produced in and through industrial growth, not selected after the fact. Growth is the pivot on which industrial geography turns, and change is the only constant in a world of persistent disequilibrium generated by the very nature of capitalist development. We begin with a quick plunge into the swift currents of economic growth – braving the many snags of growth theory – before moving on to the concrete dimension of industrialization and its geography.

Most existing treatments of urbanization, regional development, and industry location are based upon neoclassical economics and share its assumptions and shortcomings. The three fundamental building blocks of neoclassical theory are: (1) the central economic activity is exchange in price-fixing markets; (2) the goal of economic exchange is efficient resource allocation in the service of subjective preferences; and (3) the natural state of the system is to come to rest at a stable equilibrium. This is thoroughly backwards; in our view, the heart of the economy is production, the purpose of production is to generate a surplus, and the production system is in constant motion and perennially in a disequilibrium state.

Neoclassical economics begins from the realm of exchange and works back to production (Nell, 1972). It puts enormous weight on the virtues of price-fixing markets in assuring efficiency, justice, and social well being. The fundamental "economic problem" facing humankind is the scarcity of resources; the solution is to allocate resources as efficiently as possible to meet human needs. The invisible hand of the market is the surest path to this end, as households exchange their endowments of commodities, including labor and capital, to improve their lot. Those who hold these views expend great amounts of energy to proving that untrammeled exchange produces maximum satisfaction, but direct very little attention to the actual operations of production, which are represented in terms of highly generalized "production functions" that can generate steamboats as easily as tubas. Production functions join factor inputs (resources) in the right proportions to create commodity outputs in the quantities demanded by final consumers.

Our approach begins with production, the source of the commodities exchanged in markets. Production is not the passive joining of inputs but the active application of human labor, with the help of human-made tools, to natural materials in order to transform them into useful objects. In generalized commodity exchange, commodities are produced by commodities, but the key factor, wage labor, is embodied in a class of living, breathing workers. As the productive powers of human labor have expanded over the centuries, commodity exchange has expanded around the globe and market institutions have been created to facilitate these commodity flows and to coordinate disparate acts of production.

In the neoclassical scheme production is guided to beneficial ends by an exogenous set of factor supplies and consumer preferences of individuals and households. Prices of inputs and outputs are set by the intersection of aggregate supply and demand curves. Prices are key signals in the market system. They regulate individual firm and household behavior. They cannot be affected by individual action because such action is limited by perfectly free and open competition. Businesses react in a rational manner to price and profit signals, and allocate their resources so as to maximize profits, i.e., the difference between revenues and costs. Thus "profit maximization" is a curiously passive activity in the neoclassical world. Positive and negative (excess) profits call forth adjustments; zero (excess) profits signal that all is well. Profit is a *residual* that disappears in the equilibrium state to which the economy gravitates. Firms are able to equalize revenues and costs at the margin by making substitutions along their production frontier. Capital, too, is passive. It is simply a fund of homogeneous value, one of several inputs or factors of production. Capital does not make profit, but simply commands a "fair return" for its contribution to productivity.

Our view is again quite the contrary. The goal of capitalist production is to generate a surplus of output over inputs. The distribution of this surplus is at issue. Most of it goes to enrich the class that controls the means of production in the form of money profits, derived from surplus value (labor time). Capital is both money invested to make a profit and a relation of class domination and exploitation. The long-run goal of capitalists is to maximize their accumulation of capital. They are spurred on by the force of competition, which is not a matter of adjustment but of surviving and prospering by keeping up with, or bettering, one's opponents.

Production is constrained and enabled by such objective considerations as technological knowhow, installed equipment, consumer demand, labor skills, and time worked. Both profit rates and prices are themselves produced, flowing from the technical and social relations of production, not established in exchange prior to production. Prices and profit rates are, therefore, secondary variables in the analysis of production and capital accumulation – guided principally by what society can do, not what is subjectively desired. Allocation of scarce inputs is no longer the issue so much as the development and distribution of socially-generated wealth.

Finally, neoclassical economies remain at rest unless moved by exogenous disturbances. When the system does receive such a shock, rational responses to price and profit signals, perfect competition, and adjustment of production mixes assure that it will return to equilibrium. The economy grows through the expansion of factor supplies, chiefly the laborforce, and through technological change that moves production functions outward (Solow, 1970). Such change is either exogenous or induced by factor prices that trigger the search for technologies which will conserve on labor, materials, or capital.

Our view of capitalist reality is once more quite otherwise. The economy is fundamentally a disequilibrium system, driven to grow and to change by its own internal rules of surplus generation, by investment to expand capital, by fierce competition, and by technological change to extract more surplus (value) from human labor. Growth is itself produced by the systematic expansion of the forces of production, that is, through industrialization. This is an inherently creative and unpredictable process of doing things that could not be done before (and consuming things that never existed before), and cannot be uniquely determined by price and profit signals based on past or present production or preferences. Prices and profits are approximate guides in a rapidly shifting world of production. The drive to compete and accumulate constantly disrupts existing conditions of production. Equilibrium is always just out of reach; efficiency is an idea to be considered seriously when the money is running low.

Neoclassical theory actually has "no theory of growth at all" (Harris, 1978, p. 247). Growth cannot be understood without a model of how production expands over time and how the forces of change are unleashed by the economic system itself. Production is set in motion, as the discussion to follow demonstrates, by three basic forces: capital investment, strong competition and technological change. Prices, profits, and disequilibrium growth paths follow from these fundamentals of the capitalist economy.

This approach rests squarely on Marx and the classical economists, who, immersed in the Industrial Revolution, saw the principal economic questions as those involving growth, class conflict over the distribution of output, and industrial production. Today – with world industry going through yet another revolution in technology, with persistent concern about growth and national competitiveness, and with a sharp renewal of class conflict (more from above than below) in the era of Reagan and Thatcher – these questions have renewed urgency. Marx, in particular, focused on the mobilization and exploitation of labor as the first basic achievement of the capitalist industrialists, followed by the competitive search for ways to increase labor productivity through technological change, or revolutionizing the forces of production (Marx and Engels, 1848). He thereby broke with Ricardo's (pro-industrialist) concern with the struggle between industry and the landed classes over rent, and the fear that diminishing returns would stifle industrialization (Ricardo, 1821). With the long prosperity and growing conservatism of the late nineteenth century, however, conventional economics turned away from these themes to the marginal obsessions of neoclassical theory.

Schumpeter and Keynes put growth, technological change, and disequilibrium back on the agenda. Schumpeter's theory highlights competitive struggles, innovation and disequilibrium while depicting, *contra* Marx, the capitalist entrepreneur as true revolutionary . The Great Depression of the 1930s shook economics out of its torpor, stimulating both Schumpeter's finest work (1939) and the Keynesian revolution. Keynes (1936) came from a different theoretical angle, beginning with the problem of money, which had also been a fundamental concern of Marx and the classical economists. He was occupied with how the flow of payments – particularly capitalist investment and worker consumption – stimulated and was stimulated by the health of the economy. "Effective demand" (desire backed by funds) came to be seen as a critical factor in maintaining growth and staunching crises. Keynes put investment at the top of the theoretical agenda as no one had before (this contribution has often been overshadowed by questions of state management of the economy through spending and monetary regulation). He largely left matters of production to the side, however, leaving the Keynesian school poorly positioned to address questions of long-term growth and change.[1]

Ricardo, Keynes, and Schumpeter were radical by the standards of their day, and their followers have shown that their concerns are still relevant for those who would question the conventional wisdom about how capitalism works – or fails to work (e.g., Robinson, 1956; Sraffa, 1960; Mensch, 1979). We take their ideas very seriously – especially those of the neo-Ricardians about the input–output structure of production, the Schumpeterians about competitive advantage and technical change, the Keynesians about the flow of investment in a disequilibrium system – and incorporate them into our own approach. But they do not show sufficient concern for capital as a class relation, or for labor as the central element in production. Without these, one cannot capture the full dimensions of capitalist production, the revolutionizing social force of industrialization,

[1] Among the Keynesians, only Harrod (1973) made a fundamental contribution to growth theory, with his exploration of aggregate equilibrium growth paths.

nor the political dimension of growth, in the broadest sense of political economy. On these grounds, we remain firmly within the Marxist tradition.

2.1 Investment and growth

We now turn to the way capital circulates through production, and to the role of investment and profit in the expansion of production, or growth.

2.1.1 Circulation, investment, and accumulation of capital

In neoclassical economics there is only one kind of circulation: the passing to and fro of commodities in the market. Capital circulates in a different way from ordinary commodities, however: capital begins not as a commodity in search of a market but as money thrown into circulation in the hope of making a profit, that is, capital is invested (Marx, 1893). The investment of industrial capital means that employers purchase wage labor and capital goods for productive use, not for personal consumption. Production creates new commodities and these are converted to money through sale. The circulation of industrial capital thus involves three steps: investment, production, and realization. Even though the capital changes form from one step to another, there is a continuity to the circuit: ordinary commodities fall out of circulation when they are used, while capital continues through a succession of uses.

Capital investment is undertaken in order to expand the original monetary sum. While profit can be made on financial and commercial transactions, the principal means of deriving profit in an industrial economy is through production, and production generates the surplus (as product and value) on which all profit ultimately rests (Marx, 1894). Having once made a profit, however, capitalists do not stop: investing the original capital once again, plus the net profit, they can realize more profit, and so on, indefinitely. Capital is restless, always in motion; the circuit of capital is thus converted into a spiral of expansion: the accumulation of capital (Smith, 1776; Marx, 1867).

Accumulation means that the stock of capital value in its various forms – principally money, but also productive equipment and commodities for sale – becomes larger and larger with each new round of production. The pace of capital accumulation is determined by both the rate of profit on investment and the rate of circulation; the pace at which capital is invested, the time needed for production and for sales; the turnover time. Capital accumulation is the central activity of a capitalist economy, but it is not the same thing as economic growth. "Development" or "growth," two notoriously undefined terms, usually denote an increase in the productive powers of an economy, including an expanding laborforce, greater installed capacity, growing output, more trade, rising income, and so forth. These are real things, representing real wealth, and the potential for general betterment; but whether they benefit society depends on their distribution among the populace, and a host of other social considerations. Accumulation of capital, on the other hand, is measured in purely monetary (value) terms, a seemingly metaphysical, abstract gain, that nonetheless has a very real payoff in terms of corporate empires, personal wealth, and power for those who sit atop the spiral.

2.1.2 Investment and the future of production

Investment is a critical moment in the circuit of capital, as it initiates new rounds of production. The pace at which capitalists reinvest is essential to the rate of accumulation and to the speed of real growth. While bursts of consumer spending or government deficits may stimulate short upturns, sustained periods of growth are invariably led by investment booms (Abramowitz, 1976). Conversely, periods of stagnation are characterized by low levels of business investment, as in the lengthy doldrums of the 1970s and 1980s (Sweezy and Magdoff, 1988). Investment is not, however, simply a mechanical response to existing indicators. Capital can act to increase effective demand, to install new capacity, and to probe unknown possibilities in a context of change.

Investment is the capitalist's form of consumption: purchases of productive equipment (long run) and materials and wage labor (short run). Spending for investment goods is an important constituent of aggregate effective demand. Buoyant investment therefore plays a crucial role in keeping the accumulation process afloat. Without investment, demand for capital goods sags and the economy slows down, generating fewer profits. In this sense, investment creates the conditions for profitable production, or, as the Keynesians put it, "capitalists spend what they get and get what they spend" (Robinson, 1956, 1962). But there are limits to this self-propelling investment given by the underlying conditions of consumer demand, competition and technical change.

Investment is the means by which productive capacity expands and by which new techniques are installed. Plant and equipment are renewed and enlarged, and new workers and additional materials are brought on line.[2] While technological change cannot be reduced to "embodied" fixed capital alone (as some "vintage models" of capital goods assume), most work rationalization and learning-by-doing is done in concert with the development of new machines and products. Moreover, a significant

[2] We are moving between two distinct categories of investment here: fixed capital spanning several production periods, and all current outlays for production. The distinction is important in calculating profit rates and in other ways, but is not crucial to the arguments being made in this section.

percentage of new product lines and new production systems is installed in entirely new factories that require investments in land, buildings and infrastructure as well. Hence the growing quantity of fixed capital on hand is a rough and ready measure of industry's increasing productive powers (Harvey, 1982). Finally, investment in research and development and the expansion of other indirect labor activities that bear on the overall productivity and rate of innovation of extended production systems are vital to unlocking the growth potential of industrial technologies.

Investment is, finally, the principal means by which individual capitalists compete and explore the possibilities for growth in a situation fraught with uncertainty. Investors face two problems: allocating capital to the best use across a finite range of possible industries and probing the possibilities for growth within the industry(ies) where they already operate. In short, they must invest in a dynamic setting, in which some firms outperform others and some industries expand more rapidly than others. Capitalists operate, if not in the dark, then in the dusk as to the future. They have some sense of the possibilities, and can extrapolate from past rates of profit, market growth, and technical change; but they cannot predict the future. Thus investment is always an experiment, especially where it involves new methods or new products, and there is no basis for assuming perfect foresight, as in neoclassical models of rational expectations. The fundamental problem of how much to invest and in what directions is always solved imperfectly, and sometimes very badly indeed.

Capitalist investment is thus always a matter of speculating on an unknown future in a competitive environment; speculation is not something apart from the ordinary workings of the capitalist economy. Nonetheless, the speculative element can gain ascendancy over reasonable calculations of profit and loss: capital can be pushed too eagerly into an industry, overrunning the real capacity for market growth, technical change and profit making. Bandwagon effects, or mutually reinforcing expectations (or delusions), often develop, as happened with bank loans for oil exploration, real estate development, and Latin American nations in the 1970s and early 1980s (Strange, 1986). Capital can also be underinvested in potentially productive uses, or even withdrawn entirely from existing factories in a predatory "milking" process in which a company generates cash for purely speculative purposes (Bluestone and Harrison, 1982). Eventually, the wisdom of any investment must be gauged in terms of the real potential for accumulation: money cannot, in the end, simply breed money. Investment can help create the future – but not invent it out of whole cloth.[3]

[3] We cannot in this book consider the financial aspects of investment and misinvestment, including credit creation, the buildup of money as accumulation booms proceed, the rise of returns on purely financial assets beyond returns on investment for production, and the feverish financial speculation that tops off most booms and helps usher in depressions. The relation of the financial sector to the geography of capital accumulation is explored by Harvey (1982, 1985).

2.1.3 The rate of profit as a regulator of investment

As profits are both the result of investment and an incentive for reinvestment, the rate of profit plays a key role in regulating the flux of circulating capital. Capital investment is guided by the search for a satisfactory rate of profit, defined as the ratio of profits to capital invested. High profit rates signal that all is well while falling profit rates indicate dimming prospects for accumulation; high profits in general attract capital to certain sectors while low profit rates discourage investment.

Nonetheless, the profit rate must be brought down from the heights it usually occupies in economic analysis. In neoclassical economics, firms are guided by a well-behaved profit rate to establish optimal levels of output and mixes of inputs. In much Marxist thinking, as well, the rate of profit is a signal to which capitalists must and do respond very quickly, accelerating their investments, junking plants or moving to foreign climes. But this asks too much of an imperfect indicator.

In the first place, capitalists often cannot and do not respond with perfect rationality to profit signals. Aside from problems of limited information and bounded rationality, persistent difficulties arise in determining depreciation and assigning revenues and costs to particular investments where things are produced jointly and capital stock is assembled higgledy-piggledy over time. In fact, the currently accepted method of calculating the rate of profit on fixed capital was not invented until the 1920s, yet capitalists managed to function without it (Chandler, 1962). Calculated profits can also fluctuate wildly from year to year, making it necessary to work from some sort of long-run average, especially where long amortization periods or considerable advance planning are involved (Farjoun and Machover, 1983; Capoglu, 1987).

As a result, it often takes a long time for companies to grasp the full import of declining performance (as many managers have discovered to their despair in this age of "corporate raiders"). In addition the idea of an acceptable profit rate is quite elastic, the product of tradition, nationality, size of firm, and class conflict. Japanese industrialists evidently make do with a lower rate of profit, on average, than their American counterparts (Sato, 1987). This is possible because in good times capitalists do not operate at the bare margin of profitability, but ride a wave of surplus (value) that normally sweeps them high above the rocks of adversity. As a result, there is margin for institutionally- or traditionally-defined differences in acceptable profit rates, whether between groups or over time.

Most important, the signals provided by past and present profit rates, no matter how accurate, cannot foretell the future. All capital investment has an irreducible speculative element; no extrapolation can eliminate the experimental and uncertain aspect of investment. Yet future rates of profit are, to some extent, self-fulfilling prophecies in that they depend fundamentally on past and present investment; profits do not pre-exist production but are continually generated by new rounds of production and can be altered by its changing contours. There is a continual recursive interaction between investment and profit in the ongoing circulation of capital. In terms of growth theory, profit rates cannot be ascertained prior to production at all; hence the rate of investment does not tag along after the rate of profits, but helps create it. Capitalists are thus in a situation fraught simultaneously with danger and opportunity.

2.2 The two faces of competition

Competition is the search for advantage by one firm over another. It is a basic characteristic of capitalism – given its organization in units of private property, the profit motive and drive for expansion – and is a fundamental condition behind the growth of the economy. It has been portrayed in dramatically different ways, however, by opposing schools of economic thought, the neoclassical on one side and the classical, especially Marx and Schumpeter, on the other. These very different views reflect the two-sidedness of competition itself. For simplicity we shall call the former weak competition and the latter strong competition, and draw the differences broadly.[4]

2.2.1 Weak competition and its fallacies

Weak competition operates chiefly in the realm of circulation and results in adjustments toward equilibrium. Firms jostle for advantage in commodity markets: they seek market share by underpricing and by enticing consumers; they try to get the best price on materials from suppliers; and they work to secure the best labor at the lowest wage. In capital markets, they try to keep their share prices high and pay low interest rates on bonds and bank loans. Competition in markets can involve a good amount of strategic maneuvering and can bring real payoffs to firms that exploit cheaper labor or find less expensive sources of materials or

[4] On the classical tradition of competition in political economy, see Kaldor (1972), Robinson (1956), Young (1928). The principal difference between Schumpeter and Marx here is that for Marx the accumulation of capital, not competition, is the starting point for the analysis of capitalist industry. Successful accumulation causes firms to grow and come into competition with one another; competition then becomes a spur to further action (Weeks, 1981). Many Marxists reverse things by arguing that competition creates the drive to accumulate; but the social urge to accumulate precedes competition in its modern sense, and lies in private property, the appropriation of surplus value, and the general circulation and accumulation of money.

components. This is certainly not the passive acceptance of market prices and conditions depicted in neoclassical models of perfect competition and resource allocation, yet these maneuvers are limited in power compared with more fundamental alterations in production. Weak competition addresses the efficient use of resources, not the overturning of existing conditions. It is chiefly an adjustment process that attempts to move the economy toward equilibrium given a set of technological, organizational, institutional, and historical parameters.

Weak competition helps regulate the allocation of social resources by shifting labor and capital from ineffective firms and sectors toward more efficient ones. To effect this reallocation, capital must flow from lower to higher profit firms and industries. Indeed, the circulation of capital is the characteristic means for regulating social production, under the capitalist mode of production. Circulation in this sense means the inter-firm and inter-sectoral movement of investment, in addition to the inter-temporal flows discussed above. Investment flows join commodity flows (material linkages) as the fibers binding industries across the social division of labor.

One result of weak competition is a weak tendency for profit rates to equalize. The competition for (existing) capital rewards profitable firms and industries with investment funds while denying those funds to the others, causing them to run in place or even fall back. As high-profit areas attract funds, they create new capacity and new competitors; the resulting expanded output tends to lower prices, while the outlays needed to stay competitive – for larger plants, new technologies, greater marketing efforts, etc. – tend to rise. As a consequence, profit rates fall. Low-profit areas do not attract funds, so the growth of supply relative to demand is restricted; this lifts prices and keeps costs in harness. There is no assurance that profit rates will actually reach equality across all sectors and firms, however; weak competition only generates a tendency in that direction.

Economic theory has been taken down blind alleys by two common assumptions about capitalist competition: first, that competition in a market requires a large number of competitors; second, that competition between sectors for finance requires that capital be highly mobile. These fallacies of imperfect competition and perfect mobility are most zealously advocated by neoclassical theorists, but have also been widely adopted by writers from other perspectives. They are crucial to the operation of the weak form of competition, but play little role in the dynamics of strong competition.

Imperfect competition The concentration of capital in larger and larger firms, with huge market shares, is said to eclipse the classic laws of competition. This notion cuts across schools of thought, from errant neoclassicists (e.g., Bain, 1956) to institutionalists (e.g., Galbraith, 1967) to Keynesians (e.g., Kalecki, 1971) to Marxists (e.g., Baran and Sweezy,

1966). Imperfect competition does not represent a break with the vision behind the theory of perfect competition, but just an inversion of its terms.[5] The first error of these theories is to assume that in the past competition generally involved a large number of firms relative to their markets. Despite the myth of a Golden Age of "competitive capitalism," competition was historically very poorly developed especially across space. Indeed, the development of national and world markets, of mass production techniques, and of large companies operating over wide territories all of which accompanied industrialization, has greatly intensified competition. The concentration of capital is checked by the competition with other large firms from increasingly distant places, the constant creation of new companies from below (with the help of finance capital), and by changing products and process technologies.

A second error is to hold that competition is primarily about price levels. Monopoly power and collusion allow big firms to set artificially high prices, and price control is the key to excess profits. But artificially high prices and the short-term excess profits they bring are not the surest path to capital accumulation. That comes more readily from investment, efficiency, technical change, labor control, good marketing, and the like. Increasing firm size has certain advantages, but these have much more to do with economies of scale in production and marketing, labor-force mobilization and control, diversified risks and political clout than with restricted prices (Chandler, 1962). The recent difficulties of US auto companies have been an object lesson in the dangers of complacency bred by post-war conditions of market dominance and price-led profit taking.

Not surprisingly, empirical studies attempting to relate profit rates to industry concentration have made a very poor showing, with weak methodologies and abysmal regression statistics (Semmler, 1984).[6] The reason is that profit differentials cannot be explained primarily by reference to lack of competition but by the competitive process itself, and the chief motor of advantage in this process is change in the conditions of production. Competition leads capitalists to create new products, new machines, new divisions of labor, new methods of labor control and so forth; their profits vary because they do different things in different ways.

Perfect mobility In the neoclassical model, capital flows swiftly from one industry to another when an inequality of profit rates arises for any reason.

[5] There is currently a strong revolt against the theory of "monopoly capitalism" among Marxists (e.g., Shaikh, 1980; Weeks, 1981; Harvey, 1982; Semmler, 1984).

[6] Nor do wage shares or capacity utilization rates show any better relation to concentration ratios. Indeed, concentration ratios and firm size can just as easily be explained in terms of capital intensity, and recourse to collusive price manipulations by poor profit performance. In fact, small firms have profit rates as high, on average, as large firms – but they go bankrupt more often because they cannot ride out the bad times (Singh and Whittington, 1968; Marris and Wood, 1971).

Capital, like all factors in the model, is assumed to be perfectly mobile, unless exceptional circumstances intervene. As a result, profits cannot be out of equilibrium for long and no one can easily go out of business. The prevailing radical critique is that capital is mobile but labor is not; that is, corporations can move their capital at will from one place or activity to another, leaving old industries and militant or expensive laborforces for growing sectors and cheap, docile workers. The rapid "global scan" of contemporary capitalists gives them new leverage over workers and communities, by increasing the effective pool of people and places competing for the blessings of capital investment. This raises the spectre of massive deindustrialization brought about by marginal differences in profit rates (Bluestone and Harrison, 1982).

This critique is off the mark. It simply inverts the parameters of perfect mobility within a vision of weak competition and resource allocation. If money capital were perfectly mobile and productive capital perfectly adjustable, profit rate discrepancies could not persist for more than an instant. There would be no prospect of competitive advantage, no possibility of uneven development, and no problem of "capital switching" between declining and growing sectors, regions, or countries – or between old and new techniques of production (Harvey, 1985). On the contrary, profit rate inequalities are inherent to capitalist production because capital is not perfectly mobile in any of its forms. Certainly money capital moves with great swiftness and to great distances in search of more advantageous combinations of inputs and markets, laborforces and technology. Yet in order to produce surplus (value), firms must build up a productive apparatus consisting of fixed capital, workers, land, political alliances, and so forth – all with a local base. This process renders capital temporarily immobile, making it subject to some leverage by workers, communities and governments, and creating "localized" profit rates. Capital cannot adjust perfectly to the field of profit opportunities – especially as strong competition is constantly shifting the terms of production through the rise of new sectors, technological and organizational change, and the opening of new product and factor markets.[7]

Thus capital is in a bind, as Harvey (1982) has forcefully argued: mobility and immobility both offer advantages, but each has costs. If capital has become more mobile in certain respects, it has also become increasingly frozen in a landscape of immense industrial complexes, highway and rail systems, cities – and the remaking of these to suit the

[7] Even high capital mobility cannot assure profit rate equilization, it should be added, because profit signals are inadequate to the task, for reasons beyond those indicated above. In contrast to the apparently simple logic of capital flows toward higher profit regions, it has been shown that where the mass of profit is uneven, low profit rate regions can continue to grow faster in absolute terms than high profit rate ones (Webber, 1987). Also, in complex multi-sectoral production systems, where the outputs of some industries are inputs to others, it would be physically illogical to determine output levels strictly by profit rates, and equilibrating flows of capital might actually shrink higher profit industries with low profit inputs and vice versa (Nell, 1983).

ever-changing needs of accumulation is a titanic endeavor, fraught with financial and political pitfalls. Capital periodically unleashes what Schumpeter aptly called "gales of creative destruction" that install new forces of production and undo old ones that stand in the way (1942, p. 84). This process of "modernization" however, can be highly disruptive to all who live through it, including capitalists, and there are risks of failure in the form of economic or social crises. The trick, therefore, lies in maintaining a balance of mobility/immobility of capital, not just maximizing mobility.

Weak competition does not lead to equilibrium in either the short- or the long-run of economic history. Even on their own grounds, the adjustment mechanisms of weak competition do not function in ways that permit the idealized outcomes of equilibrium analysis. As Clark, Gertler and Whiteman (1986) have shown, a variety of material and institutional conditions impede the free movement of resources, while permitting capitalists generally to wield greater resource-deploying power than free-form supply and demand conditions would allow. As a result, inter-regional equilibria would not form even if technological conditions remained constant. Our main emphasis is not on the internal contradictions of the adjustment process, however. Rather, it concerns another type of competition altogether, one which defeats the adjustment process and fundamentally makes capitalist economies disequilibrium economies. We call this strong competition.

2.2.2 Strong competition

In its strong sense, competition drives capitalists to revolutionize production in order to gain an edge on competitors, thereby continually disrupting established conditions, pushing the economy to grow and keeping it from ever settling into equilibrium. Firms do not merely adjust to market conditions and keep to competitive standards; they struggle with one another in a fierce sort of economic warfare. A firm that gains competitive advantage over its rivals can earn surplus profits and increase its rate of accumulation; a firm that falls behind the others may perish. Capitalists, therefore, actively search for ways to transform business practices, to do what has never been done. This can include new products, new forms of work organization, better machinery, different materials, improved marketing, restructured administration, and creative financing. Strong competition drives industry forward and helps make capitalism a mode of production in which "all that is solid melts into air." Simply making the best use of current resources – the neoclassical vision of economic rationality – is not the point; overcoming present limits to growth is. Price signals from the past and present cannot be controlling, because revolutions in production conditions – via technological and organizational change – do not have to respect factor costs precisely (see sections 2.3 and 2.4). Disequilibrium is, from this point of view, a condition internal to capitalism, not one introduced from outside the system. Indeed,

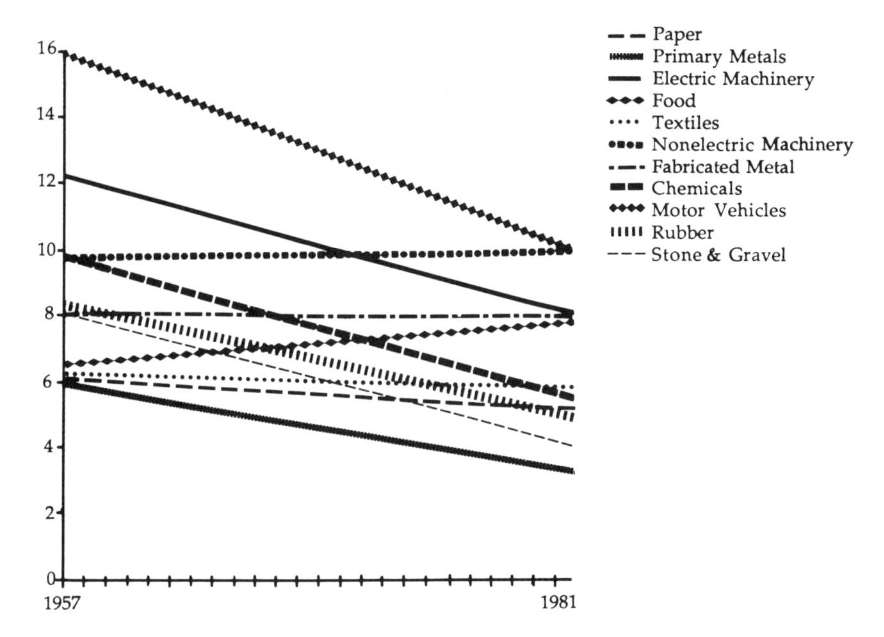

Figure 2.1 *Profit rate trends in selected US industries, 1957-1981* The diversity of profit levels among industries is readily apparent, and there is no indication of convergence over time. (After Capoglu, 1987)

equilibrium capitalism is an oxymoron, characteristic not of calm but of a state of crisis (Harvey, 1982).

In economic theory, equalization of profit rates is often treated as a fact rather than as a tendency which is never realized and which, by the nature of capitalism, cannot be realized. Equal profits are a sign that no one is innovating, that competition has come to a halt.[8] Competition, in fact, works on profit rates from both directions, and the force of strong competition systematically drives profit rates away from sectoral or national averages (Bellofiore, 1985; Farjoun and Machover, 1983).

In fact, profit rates are widely scattered across firms and industries, and there is no apparent tendency toward equalization over time. Figure 2.1 shows sectoral differences in profit rates for several US industries.

This is all very different from the neoclassical theory of perfect competition, in which the interaction between firms is not a struggle but a policing action, with everyone responding promptly and rationally to price signals set by outside conditions of changing consumer tastes, shifting

[8] It is conventional to measure profit rate differentials only within sectors, while assuming profit rate equalization across sectors (e.g., Semmler, 1984); but dynamic competition operates between sectors, and between regions and countries as well.

technologies or altered factor supplies. Firms adjust; profits equalize; competition is a gentle jostling that smooths the profit surface like so many grains of sand; capital flows like water between the grains, seeking its own level. No real competitive advantage is ever gained, the forces of production are not developed from within, and over the long run growth rates of firms and industries do not diverge systematically. As Schumpeter observed acerbically:

> The introduction of new methods of production and new commodities is hardly conceivable with perfect – and perfectly prompt – competition from the start. And this means that the bulk of what we call economic progress is incompatible with it. As a matter of fact, perfect competition is and always has been temporarily suspended whenever anything new is being introduced – automatically or by measures devised for the purpose. (1942, p. 105)

2.3 Technological change and industrialization

Technological change is almost universally considered vital to economic growth. Empirical studies of the 1950s, based on the assumption that growth consisted principally of simple augmentation of labor and capital, discovered that the "unexplained residual" accounted for around 80 percent of growth in output (Kendrick, 1961, 1973). This helped trigger a searching reappraisal of conventional growth theory (Solow, 1957, 1970) and a vigorous debate over the treatment of capital (Harcourt, 1972).[9] Yet, despite Solow's efforts, technology remains a marginal concern of neoclassical economics. As Nelson and Winter observe: "the problem [cannot] be brushed aside as involving a phenomenon [technical change] that is small relative to those that are well handled by [neoclassical] theory; rather, it relates to a phenomenon that all . . . acknowledge is the central one of economic growth. The tail now wags the dog" (1982, p. 205). One must therefore look to other theoretical traditions for an adequate conception of technological advances in the economy.

2.3.1 Technological advance and economic growth

Technology refers to the general capabilities of human society to transform nature into useful products for human consumption, and this takes four main forms: as general theoretical and practical understanding of how to do things (social knowledge), as objects (goods), as installed techniques

[9] The theories of "human capital" (Becker, 1964) and "learning-by-doing" (Arrow, 1962), and studies of the origins of technical change in market demand (Schmookler, 1966) and research and development (Mansfield, 1968), were all stimulated by the empirical puzzle of the unexplained residual (cf. Nelson and Winter, 1982).

of production (processes), and as the personal know-how and abilities of workers (skills). Technological change occurs in all these dimensions. (We take up these issues further in chapters 4 and 5.)

Technological change generates economic growth by altering both processes and products. Process alterations are generally intended to increase the output generated by a given quantity of inputs, product alterations to increase the range and quality of goods and services available. The two overlap where products of one industry are involved in the production process of another, and, to a lesser extent, when consumer goods improve the health, mobility, and so forth of the workforce. In the long run, industrialization has greatly increased the average productivity of labor, reduced the quantities of natural resources consumed per unit of output, created millions of novel products, and vastly multiplied the number of machines. It has, at the same time, expanded the volume of social surplus (value) available for the disposal of the capitalist class to use as investment or for personal consumption.[10]

Machinery and other technical improvements have, above all, raised the average productivity of human effort. In Marxian terms, the rate of surplus value has increased, and commodities have been cheapened (the average value of each unit of output has been reduced). The virtue of putting the matter in labor-value terms is that human beings are the central force in the production process. "Production of commodities by means of commodities" (Sraffa, 1960) does not capture this fact satisfactorily (Harris, 1978). Labor is the key "factor" – the *differens specifica* – distinguishing industry from natural processes (Marx, 1867). Mechanical advance can proceed for a time without direct reference to labor, but production as a whole cannot: ultimately, as technological change proceeds, there must be revolutions in labor processes, in labor relations, and labor market practices (Walker, 1988b).

On the other hand, Marxists have in some ways underestimated technical change. They often fail to recognize that industrialization has also raised the productivity of virtually every input. In downplaying the physical side of production, they have frequently missed crucial differences among technologies and glossed over the heterogeneity of industries in much the same manner as the neoclassicals. Furthermore, product innovation has been underplayed in comparison to the labor-saving aspects of mechanization (e. g., Shaikh, 1978, 1982; Sayer, 1985). From the dawn of modern industry, product innovation has been fundamental, as such nineteenth-century introductions as railways, coal-tar dyes, kerosene, arc lights, steel, elevators, sewing machines, revolvers, and steamboats attest

[10] Increasing productivity does not apply to capital, which is a social relation pertaining to class ownership, distribution of the social product and the circulation of value. As the Marxian, Keynesian, and Sraffan schools of economics have shown in various ways, capital is not a "factor of production" in the conventional sense.

(Schumpeter, 1939). Why would capitalism tend to generate new products? If the firm produces a better version of an existing product, it drives out competitors still producing inferior versions; it gains market share and surplus value with increased output of the same basic product. With fixed consumer buying power, the new simply knocks the old from the set of goods purchased. But new products are, in fact, associated with the process of growth. If a new product finds a new market, it means the capitalists can employ additional labor, adding a new quantum of surplus value to the larder of the firm.

Neoclassical theory, on the other hand, makes it appear as if all new uses of capital and labor – or, put more broadly, all new product lines or ways of making products – represent tradeoffs against existing or alternative uses of inputs, as captured in the notion of "opportunity cost." This kind of resource substitution may be sensible in a static setting, but makes a fundamentally unrealistic assumption about technological change and growth. According to the law of diminishing returns, the marginal contribution of adding more units of any input declines at a rate described by that activity's production function. Efficient production takes place at the point where the relative marginal productivities of capital and labor match their relative prices. Diminishing returns are required for the construction of well-behaved neoclassical production functions. But diminishing returns are a poor assumption in a technologically dynamic setting.

Industrial production can, over time, generate more output with fewer total inputs. This is conventionally called "increasing returns to scale," which is a somewhat misleading term. While returns to a given quantity of inputs are increasing, and the scale of output is expanding over time, the latter is less the cause than the effect of the former. Real increasing returns to scale appear only in certain limited cases: where the size (area, mass, etc.) of an object increases exponentially to its exterior dimensions (e.g., the volume of a storage tank to its height and diameter); where the average unit cost of a commodity declines with the number of users (e.g., a large bridge is not economical to build for a single user); where continual use of a machine over time reduces average costs (e.g., running a mainframe on a 24-hour basis); and so on. But the essence of technological advance is not restricted to such effects.

Increasing returns over time are even more dramatic than returns to scale. By this we mean the ability of new technologies to do more with less, or to create previously unimagined kinds and quantities of products from the same humble capacities of the earth and human labor. This simple fact is devastating to the neoclassical model of perfect competition, profit equalization and general equilibrium, and demands a completely different theoretical approach to understanding economic dynamics. Realizing the dilemma, Marshall (1900) tried valiantly to skirt the problem by accommodating both increasing and decreasing returns to scale within the same model, making increasing returns a special case, like imperfect competition. Marshall's effort was annulled by Sraffa's (1926) critique of the laws of returns, but the category has taken on a life of its own. Efforts to overcome the problem internally, as in Kaldor's (1972) theory of "dynamic economies of scale," are probably less helpful than returning to the older concept of industrialization. Technological change is simply the microeconomic term for industrialization, or the continual development of the forces of production that has been going on at a brisk clip since the eighteenth century (Marx, 1867; Mantoux, 1961; Landes, 1970). Over time, industrialization produces more with the same endowments; technological change is a principal means by which it does this. As an interdependent set of events working together to transform the productivity of labor, virtually all technological changes can be interpreted as contributing to increasing returns to industrial activity. Indeed, the Industrial Revolution and its modern history have given the lie to the pessimism of Malthus and Ricardo about diminishing returns in the long term, even in agriculture.

The existence and nature of increasing returns in industrial production is highly contested, both theoretically and empirically (Cripps and Tarling, 1973; Rowthorn, 1975; Thirlwall, 1980; McCombie, 1981). We cannot do full justice to these controversies here, but much of the skepticism concerning the existence of increasing returns is founded on an (incorrect) distinction between changes in production techniques and increasing returns "without changes in technique." If increasing returns attach only to an existing production function, while any other changes in returns are pushed off to a separate category of technical progress, the analysis becomes static, and increasing returns are largely excluded by definition (Storper, 1988). Furthermore, where empirical studies inescapably show increasing returns over time, economists typically assume that the changes in technique which generate them are purely internal to individual firms, excluding the combined effects of many firms interacting with one another. Yet it is clear that increasing returns are not confined to particular firms; industrialization is, above all, a process in which gains in productivity are created in the form of increasing external, as well as internal, economies of scale.

External economies cannot be separated from technological progress because such progress depends on the development of the forces of production as a whole, including increasing specialization and interconnections between functions in the economy. Social production has become progressively more elaborate, both by subdividing tasks within a given factory (the detail division of labor) and by increasing specialization of activities among workplaces, firms, and branches of industry (the social division of labor) (Smith, 1776; Marx, 1867). The expanding division of labor contributes to productivity growth through the acquisition of special skills and development of specific machinery as well as by simple economies of scale and scope in the use of equipment and labor.

Young (1928) dubbed this the increasing "roundaboutness" of social production, observing that, "The mechanism of increasing returns is not to be discerned adequately by observing the effects of variation in the size of the individual firm or of a particular industry . . . [W]hat is required is that industrial operations be seen as an integrated whole" (p. 539). Increasing returns are not strictly limited to a particular firm or workplace because products, machines, and labor processes are always developed in the dynamic context of the economy as a whole.

The splitting off of new activities in the social division of labor may be thought of as embodying external returns, insofar as specialization occurs between firms or industries, but the same lesson holds for the operations within a factory: increasing task specialization may reduce the scale of an individual unit of operation even as this subdivision of tasks means an increase in the scale of operations of the factory as a whole. The returns to scale are external to the individual task within the detail division of labor. In a more dynamic sense, this interdependence means that "[e]very important advance in the organization of production alters the conditions of industrial activity and initiates responses. . .which in turn have a further unsettling effect. This change becomes progressive and propagates itself in a cumulative way" (Young, 1928, p. 533). In reality, "tasks," "firms," and "industries" are constantly being redefined, so the notion of returns as either strictly internal or external to firms loses theoretical significance. Shifts in production techniques are simultaneous and multifarious. Industrialization is a process of unevenly developing roundaboutness in production, over which no single agent has control and whose very interdependence (and thus the external nature of increasing efficiencies) precludes the possibility of developing equilibrium allocations of resources.

In neoclassical theory, the "economic problem" is identified with the allocation of scarce resources, but technological change and industrial growth are fundamentally about expanding the resource system of the economy. With increasing returns and expanded reproduction, every resource contributes as much or more output in its actual use as it could in any alternative use. The "pattern of use of resources at any time is no more than a link in the chain of an unending sequence and the very distinction, vital to equilibrium economics, between resource creation and resource allocation, loses its validity" (Kaldor, 1972, p. 1245). Technical change must therefore be treated within a completely different axiomatic system of competition, investment and the logic of technological paths of development.

2.3.2 Price-induced technological change

If the allocation of scarce resources were the guiding principle then technologies – which embody distinct combinations of inputs and outputs – would be adopted in conformance with price signals set by market conditions of supply and demand for economic goods. But this neoclassical theory of price-induced technological change is not viable because of the dynamic economies inherent in technological progress and in industrialization. To a certain degree technological change is "market induced." Given the information and alternatives available, capitalist managers attempt to respond to market signals. But a neoclassical world requires perfect knowledge, smooth substitution, and a variety of more or less equivalent techniques, and these unrealistic assumptions do not accord with the technical nature of production and its development (Rosenberg, 1976, 1982).

Neoclassical production functions are constructed on the assumption that a set of alternative production techniques is readily available for every product. This allows substitution of one input for another as their relative prices change, or what is usually called "choice of technique" and "factor substitution." These assumptions have been widely criticized (see e.g., Harcourt, 1972; Hunt and Schwartz, 1972). For one thing, the number of techniques actually available is usually quite limited, so movement from one to another tends to be jumpy. Commitment to fixed capital which cannot be instantly abandoned or converted contributes to the rigidity of technique: capital is not putty, as Robinson (1962) has observed, and cannot be squeezed into whatever shape currently best suits the employer. These criticisms challenge the theoretical requirement for continuous adjustment along the production function, but they are rather like the theory of imperfect competition in treating the problem in undynamic terms, i.e., as a weak competition among techniques exhibiting diminishing returns. A different sort of challenge has been entered by the "reswitching" paradox put forward by Sraffa and his followers (Sraffa, 1960; Pasinetti, 1977). By introducing cross-sectoral inputs, they demonstrate that the choice of technique can be perverse as the wage-profit ratio changes; that is, as wages rise, capitalists in one sector will adopt labor-saving techniques for a time but suddenly switch over to capital-saving techniques as the wage rate passes a certain threshold (and back again later). This helped undermine the neoclassical edifice, but took the debate down the wrong road by ignoring the more fundamental problem of growth and technical progress (Robinson, 1979; Pasinetti, 1981).

The essence of technological change is increasing returns. The choice is not between roughly equivalent techniques with different factor ratios, the choice is between present and future techniques as industrialization moves forward. New techniques outperform existing ones in significant ways, and often across the board – i.e., they are not just "labor saving" or "capital saving," but save on all inputs at the same time. In the neoclassical vision, competing technologies show diminishing returns, and market sharing is the result. The outcome is completely predictable – we can determine in advance the market shares of each technique – and the configuration that emerges is demonstrably efficient, flexible and well-behaved. Increasing returns undo this model because the economy grows right past existing price signals, as it were. Hence, outcomes may appear to be perverse because firms need not conserve on the most expensive input if overall productivity goes up enough to make their choice of technique profitable and competitive anyway. For example, a metal fabricating firm is hit by a sudden price rise for materials, which triggers a search for a way to restore profits. It discovers that new types of labor-saving machinery have come on the market, but that the machinery is less tolerant of odd-shaped pieces which the firm had used to stretch its materials; as a result, the materials intensity of production increases. But the firm is still better off than before (Nelson and Winter, 1982).

Once the future is brought into play in the present choice of techniques, substitution theory collapses completely. Neoclassical theorists have tried to get around this by introducing a measure of present control over future techniques. Solow (1957) proposed that technical change be viewed as a shift outward in the production function, rather than as substitution along a static frontier. Factor-saving techniques are introduced as input prices rise over time, and the direction of change varies with the relative weight of factor scarcities (Schmookler, 1966; Hayami and Ruttan, 1971). Substitution analysis is thus preserved by extending it to the substitution of future for present technologies. The creation of future technologies is under the guidance of firms, through a separate process of investment in research and development (Mansfield, 1968) or installation of fixed capital (vintage models) (Solow et al., 1966; Salter, 1966). Nonetheless, as we shall show in chapters to come, technological change is the result of the interplay between the development of knowledge, economic necessity, and myriad cultural and noneconomic factors; thus the future path of technical change can neither be known beforehand nor entirely shaped by conscious choice, so present-day prices cannot control the direction technology takes.

Even if one has full knowledge of all available techniques, outcomes are usually indeterminate where techniques with increasing returns compete. There are, in neoclassical terms, multiple equilibria. Adoption of any one such technique is preferable to staying put, even if the one chosen is not optimal. Nor is the choice predictable, since profit maximization cannot be assured; under strong competition simple advantage is a sufficient condition for adoption. A classic hypothetical case involves introducing cars to an island without traffic laws: drivers are free to choose between the right- and left-hand sides of the road. As more drivers choose one or the other, the rest very rapidly fall in line because of the obvious disadvantages of bucking the trend. But the outcome is indeterminate. Collective action is not the sole reason for such effects, of course; they may derive from the cumulative impact of learning-by-doing, or factor creation and attraction by the winning side, or from other causes.

"Superiority itself becomes a function of adoption or use" (Arthur, 1989, p. 25).[11]

A more useful concept than choice of techniques, then, is "search for new techniques" (Nelson and Winter, 1982). The uncertainty of the search is amplified by the impossibility of knowing either the full range of available techniques or the future course of technological change. Moreover, firms, industries and even nations can become locked into particular paths of development by past choices. On the hypothetical island, choosing right- or left-handed driving gives rise to a whole paraphernalia, from car design to traffic light placement to rules of the road. Inertia, learning-by-doing and learning-by-using, the accumulation of technical artifacts, the creation of institutional frameworks of action, the weight of collective action, shutting out less immediately conceivable alternatives – all contribute to following a once-chosen path of development. For instance, the United States in the nineteenth century was noted for rapid development of machinery for tasks still carried out by hand in Britain. A neoclassical explanation is that US mechanization was triggered by relatively high wages (interest rates were higher still) (Habakkuk, 1962). Against this, David (1975) has argued for a cumulative process in which early technological innovations became the basis for later ones. If rapid adoption of machinery raised the productivity of labor, then wages were likely to rise over time and interest rates to fall because of ample profits. Indeed, in the long run it is clear that the high standard of living in developed capitalist countries is the product of the Industrial Revolution, not the reverse; capitalism has unleashed a process of rapid technological change and growth that has allowed absolute wages and consumption levels to expand without threatening profits and accumulation (Kaldor, 1956). Neoclassical theory would have us believe that technical progress is triggered by rising factor prices alone. But this is plainly contradicted by the many instances of short- to medium-term mechanization without increases in real wages, as in England in the early nineteenth century or Brazil today (Deane, 1965; DeJanvry, 1981).

Obviously capitalists try to weigh their revenues and costs in making production choices in the short run. One does not buy an expensive machine to handle a job any worker can do; relative input prices do limit action. The availability of cheap labor may retard the adoption of machinery at particular moments, and expensive, militant labor may propel the adoption of

[11] This means that where the market share of each technology follows a stochastic process – a random walk – and the menu of technologies shows standard diminishing (or constant) returns, the random walk has reflecting barriers. The aggregate outcome then may logically be efficient, flexible, predictable and ergodic. But if technologies show increasing returns, the same random walk has absorbing barriers, in which case the outcome is: (1) non-ergodic – small events at the outset are not averaged out and forgotten but may decide the path of market shares; (2) potentially inefficient – even where individual choices are rational; (3) potentially inflexible – in that ultimate market shares cannot always be influenced once development has gone beyond a certain point; (4) non-predictable – in that knowledge of supply and demand does not suffice to predict market shares (Arthur, 1989).

new machines. "It would be quite possible to write a history of the inventions made since 1830 for the sole purpose of supplying capital with weapons against the revolts of the working class" (Marx, 1867, p. 436). Nonetheless, these considerations are secondary to the drive to accumulate by producing relative surplus value on the basis of developing the forces of production. Marx railed against the "shameful squandering of human labor power for the most despicable purposes . . . in England, the land of machinery," but he understood that England's Industrial Revolution had occurred despite the cheapness of labor power, and that the industrial surplus population was, in part, a product of mechanization itself (Marx, 1867, pp. 393–4).

In short, the relative prices (income shares) of capital and labor can shift back and forth over time, but mechanization is essentially a one-way process of technological advance. While older machinery is often used in cheap labor countries, no industry ever demechanizes significantly in the face of lower labor costs.[12] The classical economists have thus captured the more basic process of development, the neoclassicals a more restricted arena of behavior.

2.4 Price and profit in a dynamic setting

Having demoted prices and profit rates to a secondary position in guiding capitalist action, we can now examine the actual mode of price and profit formation in a growing economy. Prices are actively constructed by conditions of production, as set in motion by the three principal strands of a viable growth model: investment, competitive strategy, and technological change.

2.4.1 Prices and production

In classical economics, market prices for manufactured goods were seen as oscillating around long-run centers of gravity whose primary determinants are the costs of production. Smith (1776) called these "natural prices," while Marx (1894) used the term "prices of production." Consumer demand plays a secondary role in this formulation: it affects the level of output but subjective preferences cannot determine the material cost of producing commodities.[13] Classical economics provides for the *regulation*

[12] Why adopt an inferior technique where inputs are cheap, when high productivity and low costs can be a double source of profit? One often finds the most advanced, capital-intensive production technologies being used in conditions of extremely low wages in the Third World. Where more primitive techniques are employed it is usually due less to low wages than to lack of capital, inability to operate sophisticated machinery, and other structural, institutional and political conditions.

[13] The prices of non-produced or truly scarce items, such as land or artwork, involve a measure of rent, which is treated as a separate problem in classical theory.

of market prices, rather than price *determination*, by production costs. Market prices are not equilibrium prices (Semmler, 1984); actual prices are equal to prices of production only by accident. It is not difficult to see why. Market prices are subject to short-term fluctuations in supply and demand, to speculation, and to restrictions on the mobility of capital (Clark et al., 1986). The center of gravity, on the other hand, changes only as the average of different firms' production techniques and input costs change. In the world of real space and time, there are neither infinite elasticities nor perfect inelasticities of supply and demand.

Production involves specific quantities of labor, materials and machines brought together to generate definite quantities of outputs. Every production unit in the division of labor has a characteristic "input–output coefficient," or technical combination of inputs and outputs. Complex production can be thought of as consisting of a sequence of steps in a chain, in which each step uses inputs that are products of earlier steps, or as a system of simultaneous acts of production, linked together in an input–output matrix. The classical political economists believed that one could reduce the prices of all commodities to quantities of a single input, labor, by tracing the long chain of inputs back to its origin ("dated labor") or by solving a set of simultaneous equations ("transforming simple labor values into prices of production"). The so-called "transformation problem" in Marxian value theory has, in particular, been a bone of contention (Steedman et al., 1981; Mandel and Freeman, 1984).

The crucial difficulty turns on two things: every step or branch of production uses different ratios of labor and non-labor inputs, yet all non-labor inputs provided by capitalist investment must earn a profit rate, which is normally assumed to be equal across all investments (wage rates are also assumed to be equal throughout). In Ricardian terms this means that profits and wages enter differentially into the prices of commodities; in Marxian terms, it means that surplus value must be transferred from sectors using above average quantities of labor to those using above average quantities of capital. Everyone stumbled on the mathematics of the problem until Sraffa (1960) came up with the brilliant device of the "Standard Commodity" (an invariant numeraire between the input–output equation and the price equation); it is now understood to be a matter of ordinary matrix algebra (Pasinetti, 1977; Harris, 1978). Yet debate still rages over the economic meaning of the mathematics. The Sraffran, or neo-Ricardian, school claims that labor value is a redundant category once the technical input–output coefficients of production are given and the wage (or profit) rate is known (Steedman, 1977). Marxists reply that labor is the central force of social production and the exploitation of labor-power the core social relation of capitalism, and this cannot be reduced to a technical process to which wages and profits are added (Shaikh, 1982, 1984; Farjoun and Machover, 1983). This contentious issue cannot be settled here. Unfortunately, it has obscured the large area of agreement between the Marxian and neo-Ricardian schools, i.e., that prices rest fundamentally on the structure of production and class relations of distribution, and that these are, in turn, embodied in the long-run process of capital accumulation and development of the forces and relations of production.[14]

The standard neo-Ricardian formulation for prices of production runs into trouble once some of its basic assumptions are dropped, however. First, it tacitly incorporates competition only in its weak, allocative role (Shaikh, 1980) and assumes a complete equalization of profit rates rather than a *tendency* to equalization. Strong competition, recall, is as much a force for differentiation of production conditions – and hence profit rates – as it is for their equalization. With unequal profits there is no longer a determinate solution to the system of simultaneous equations conventionally used to reveal prices of production (Farjoun and Machover, 1983).

Standard neo-Ricardian models also take technical coefficients as given, when they are, in fact, in continuous flux. Competition and the drive to accumulate render every commodity, every machine, every measure of output transient. If input–output coefficients are not stable owing to technical change, a fierce "index problem" is created that not only wrecks the ordinary solutions to systems of equations, but also makes those solutions – were they achievable – meaningless in any real sense, for the basis of price regulation is changing. Pasinetti (1981) has shown how to solve both these conundrums mathematically, but has in the process completely transformed the problem to one concerned with the dynamics of price formation with uneven technical change across sectors (Walker, 1988a). His contribution puts classical theory on a new footing (Harris, 1982).

In short, the price and value question must be posed in different terms than those of the simultaneous equation models. The classical economists, particularly Marx, were not principally concerned with the measure of commodity values and prices, but with establishing some first principles of accounting before getting on to the more profound analysis of industrialization and class conflict (Robinson, 1977; Bellofiore, 1985). The price equations must be set in motion.

2.4.2 Industry growth and profit rate differentials

The persistent differences among profit rates are not accidental disturbances to equilibrium, but essential conditions of a dynamic capitalist

[14] Unlike Steedman, Sraffa and Pasinetti have never entered into the jousts, and both have insisted on pursuing the "chimera" of reducing prices to dated labor. Robinson has observed that "Piero [Sraffa] has always stuck close to pure, unadulterated Marx and views my amendments with suspicion" (1977, p. 56).

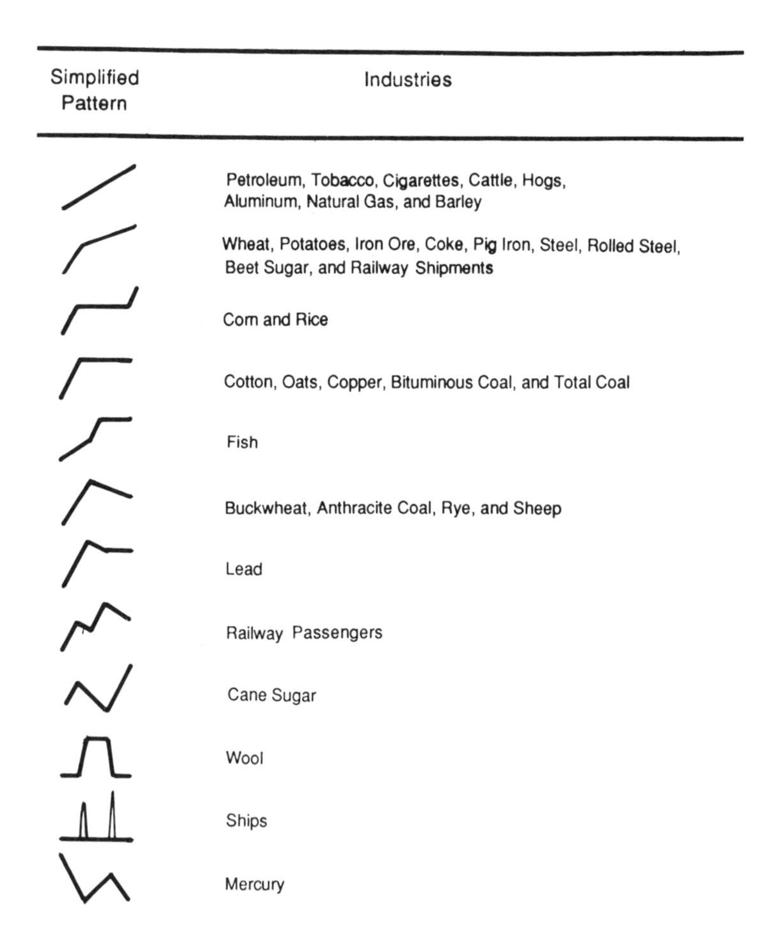

Figure 2.2 *Simplified representations of industry growth paths* The variety of patterns described by movements in output is striking, and there is no dominant tendency, such as an "S"-shaped curve. The curves are based on levels of output over various periods from the late-nineteenth to mid-twentieth centuries. For a detailed explanation, see Gold (1964). (After Gold, 1964)

economy. Strong competition and technical change are forever generating surplus profits for the fortunate and sagging profits for the laggards. At the firm level, competitive advantage promises the leading firm a period of surplus profits (or quasi-rents on its proprietary technology) relative to the industry average (Schumpeter, 1939; Mandel, 1975; Markusen, 1985). This differs completely from the neoclassical view, and from equilibrium Marxist models, in which technologies with indivisibilities create below-normal transitional profit rates (Okishio, 1961; Roemer, 1980).

In the disequilibrium view of growth, superprofits are a major motivation to competitors searching for new product and process technologies, and a source of divergent cumulative rates of growth among firms. Even if they start at the same place, the growth paths (and sizes) of firms must eventually diverge by a wide margin (Nelson and Winter, 1982).

Divergent average profit and growth rates among industries are common as well, but insufficient attention has been given to the significance of these differences in aggregate models of economic growth (but see Steindl, 1952; Kalecki, 1954). Industries differ systematically in profit rates, growth of output, rates of growth in productivity, rate of technical innovation, and size of research and development effort (Burns, 1934; Chenery, 1960; Kendrick, 1961; Mansfield, 1981). These differences rest, above all, on the fact that industries make different things, using different raw materials, machinery and labor processes; that is, they rest on different technical foundations. In short, industries develop along divergent trajectories, for reasons we shall explore at some length. An idea of these trajectories can be gained from the profit rate and output patterns depicted in figures 2.1 and 2.2. Every conceivable pattern of output growth has been followed by one industry or another: steady rise, upswing followed by leveling off, growth to a peak and then decline, rise and fall followed by renewal, and so forth (Gold, 1964).[15]

Fast-growing industries expand because their products open up whole new worlds of consumption and/or become widely affordable because process innovations reduce costs. This was true of steel in the 1880s, automobiles in the 1920s, chemicals in the 1950s, and microelectronics in the 1980s. Industry-wide superprofits are possible in favored sectors because customers are willing to pay market prices above the price of production of the good. In the taut world of neoclassical substitution theory, incomes are stretched to the breaking point but most households and businesses do not operate on the margin at all; they have surplus of income above need (or basic efficiency) that may be directed toward newly-desired ends without the short-term tradeoffs implied by substitution theory. Buyers are willing to pay a premium for commodities that promise a special payoff in productivity, competitive advantage or fashion, as in the cases of sewing machines, automobiles and personal computers in their heyday. Conversely, where process changes are causing the price of a desired product to fall rapidly – often the case in dynamic new or restructured industries such as plastics in the 1950s or electricity in the 1920s – it is all the easier to capture excess profit through generous price margins without discouraging consumers. Growing industries have always, to a large extent, created their own markets with attractive new products and falling prices. In addition, as products such as the elevator or

[15] A similar cacophony of histories can be found at a finer level of disaggregation to individual products (Polli and Cook, 1969).

automobile become integrated into the production processes and products of other industries, or into the infrastructure of everyday life, their elasticity of demand decreases, allowing prices to remain elevated for sustained periods (cf. Koutsoyiannis, 1984).[16]

There are, by contrast, many low-profit sectors in the economy. For example, in clothing, the nature of the product – its odd shapes and uncertain (i.e., fashion-oriented) market – impedes technological change, promotes high degrees of product differentiation, and traps producers in a permanent cycle of very high levels of market contestation. Thus producers are forced to adopt short-term cost-reduction strategies at the expense of long-term investment, a choice that perpetuates the existing technical stagnation and low barriers to entry. As a result, many of the firms in this industry are perpetually at the low end of the profitability spectrum.

It may be protested that labor costs are not adequately incorporated into this model of industry growth. We do not mean to imply that labor supplies, wage rates and employment relations are insignificant to the well-being of industries, a point explored in depth later. Nonetheless, here, as in the aggregate, labor costs are not the principal variable determining the rate of accumulation in developed capitalist economies (Marx, 1867; Robinson, 1977). Growing industries can afford to pay their workers (or key segments of the workforce) well, by sharing some of the excess profits brought by rapid accumulation: this has happened repeatedly, from Singer's machinists to Ford's assembly line workers to Silicon Valley engineers. High wages are not a barrier to growth in such cases. Conversely, wages in sectors such as garments today are limited by low rates of growth and low profit rates.[17]

In short, there are systematic and persistent forces at work setting the parameters of growth and the rate of accumulation across the range of industries. These make rapidly-growing industries different from slow-growing (or declining) ones in fundamental ways: they have the market power to attract resources (factor inputs) and create their own environment, and this, as we shall see, has a profound effect on their ability to develop cities and regions.

2.4.3 *Prices and profits, investment and growth*

We have seen that industries are systematically differentiated at any given moment by their underlying growth potentials and profit rates, and that

[16] Aggregate effective demand will also be a factor in the sustainability of growth, and this will depend heavily on the state of the wage bargain drawn up between capital and labor across industries (see chapter 8).

[17] Many successful industries, firms and countries combine high rates of innovation with scandalously low wages to achieve doubly high rates of accumulation. A good example is California's agricultural sector.

profit rates are associated with a certain flexibility in prices and consumer behavior. But the connection between growth rates and price formation has not been elucidated. We need to add a consideration of long-run expansion of production to the classical theory of price formation, putting it ahead of profit rates in the calculus. But we also want to incorporate the Keynesian insight that investment is the key force which activates growth possibilities.[18] The rules that capitalists follow in carrying out investment are intimately related to differential profits, and therefore investment patterns are also highly differentiated by industry.

Capitalists establish investment targets that they think will allow their firms to grow at a satisfactory rate by penetrating new markets, upgrading equipment, expanding capacity, introducing new products, or a host of other strategies. They adjust their rates of investment in light of past success and estimates of future possibilities (Jorgenson, 1971; Clark, 1979). Investment requires capital funds and, over the long run, such funds are raised from internally-generated profits, whether capital is self-supplied or borrowed and repaid with interest. Firms therefore establish "profit targets" suitable to their investment plans. Prices are then set as a mark-up over average unit cost so the desired flow of profits and investment can be maintained over time (Capoglu, 1987). Setting high investment targets allows expanding sectors to generate funds suitable to their promise, while low investment targets and low prices in declining sectors undercut competitors and bar external investors. This model accords well with what is known about business behaviour – virtually all companies employ mark-ups. These mark-ups are not principally a result of monopoly power, however (Kalecki, 1971). Rather, they are a practical method of allocating capital and generating funds.[19] Empirical estimates of prices as a function of unit costs plus unit profits (explained in terms of investment levels), show an extremely good statistical fit (Capoglu, 1987). More broadly, this model of pricing rests on underlying rates of growth rather than rates of profit: "What the systematic and persistent forces determine, then, is the rate of accumulation; competitive mark-ups will then establish a rate of profit. All that is necessary, in this view, is to redefine the 'long-period method' [of economic analysis] with the rate of growth playing the role formerly assigned to the rate of profit" (Nell, 1983, p. 115).

[18] This does not mean that we grant investment the virtual autonomy from production conditions it reaches in some Keynesian writings (e.g., Robinson, 1962, pp. 82–3; for an overview see Harris, 1978, ch. 8).

[19] Obviously, there are limits to the amount of investment funds generated through higher prices alone, but we cannot take up here the role of credit and financial institutions. Faster-growing industries and firms borrow more, but the differences are significantly less than the relative variations in investment rates. In the United States, industrial corporations have long preferred to generate funds internally, and the ratio of investments over internal funds (net of stock dividend payments) is close to 1.0 across all industries and over time. In Japan and Germany, on the other hand, external funding by banks is much greater, owing to long standing differences in financial practices (Capoglu, 1987).

The relationships between prices of production, profit rates, and investment targets may now be summarized. Capitalists attempt to set mark-up prices that will cover costs of production and allow them to realize their investment targets; these targets are set so as to attain a satisfactory medium- or long-run rate of capital accumulation for the firm (or industry) given a set of sectorally-specific technological conditions and prospects. As these prices are based on long-run conditions of growth, they may be called "prices of expanded reproduction," and they operate as centers of gravity in the same way as prices of production in the ordinary classical model. Such prices incorporate the long-run allocation of capital to investment in new capacity, replacement of worn out machinery, research and development of new technologies, and so on, as needed for sustained expansion. This is very different from a vision of price-setting based on equalized profits and static production technologies, for current prices now incorporate a logic of ever-changing technologies and capitalists' attempts to maximize rates of accumulation. In Marxian terms it means the reallocation of surplus value from sector to sector based on the amounts of capital invested for future expansion, not just the proportion of capital goods in current production. In neo-Ricardian terms, it means factoring the rate of growth, as well as the rates of profit and wages, into the price equations (Pasinetti, 1981; Walker, 1988a).

The concept of prices of reproduction is an idealization, of course. In reality, as capitalists probe the future with investments and compete against others whose strategies are unknown, the collective effects of individual actions may be unsustainable for the system as a whole. Disequilibrium growth means that uncertainty plagues every action of the individual capitalist. Equilibrium models assume that competition is set into motion after production techniques are known and price and profit levels are set. But future conditions of production are never known, so investment – and hence prices and profits – always contains an element of speculation. One speculates on a technical innovation and, if successful, alters the technological framework of production and the course of future technological change.

Today's prices, then, are a product of complex determinations, involving different structural constraints than those considered in neoclassical thinking (technology is more important than supply and demand, for example), and at the same time offering different types of strategic possibilities (different production price/market price gaps and differential prospects for technological change, for instance). In this sense, prices and profits are less autonomous than in neoclassical theory: they are not the progenitors of the form of production and the quantity of growth. While they serve as useful regulators of commodity circulation in the market, they are also important vehicles for the strategic actions of capitalists in pursuit of competitive advantage and rapid accumulation, actions which are steered more fundamentally by technical change, changing consumption patterns, and class struggles over work and income.

2.5 The rhythms of growth

We have thus far set forth a model of capitalist growth based on three essentials: accumulation (investment), strong competition, and technological change (with divergent industry growth paths). The value of this has been to stress growth and development over static allocation of resources in the process of industrialization. We are principally interested in the role of such industrial dynamism in explicating the inconstant geography of capitalism. Nonetheless, it would be wrong to leave the impression that capitalist industrialization generates only growth and prosperity. Two centuries of capitalist development show two patterns clearly. One is the long secular rise in output and productivity; the other is periodic recessions or depressions. Capitalism is beset by temporal disequilibrium, appearing as alternative booms and slumps. The rhythms of growth describe a jagged upward path.[20] These rhythms, too, rest in large part on the meshing of the three basic elements of capitalist growth outlined here.

Both Schumpeter and Marx provide critical insights into the generation of growth under conditions of disequilibrium. Each had certain blind spots but these can be overcome by selectively wedding their visions. Schumpeter was overly optimistic about the generative powers of technology because he did not account for the contradictions of the capitalist investment process; Marx advanced a theory of capitalist crisis which underestimated the dynamism of technological change.

For Schumpeter (1934, 1939) technological change is the prime motor of capitalism. Clusters of innovations appear on the scene and spur new investments and hence expansion. Cyclical downturns in capitalist economies come when the possibilities inherent in these innovations are exhausted. A very literal reading of Schumpeter holds that basic innovations ought to be clustered in periods of depression (Mensch, 1979), but the evidence does not show either clusters of basic innovations just prior to growth spurts or the exhaustion of technological possibilities just prior to downturns (Freeman et al., 1982; Van Duijn, 1983). Two lessons may be drawn from this failure. One is that the empirical work has attempted too close a temporal correlation between innovation and growth

[20] The empirical evidence reveals four types of business cycle, of varying periodicity: 3–5 years, 7–11 years, 12–20 years, and roughly 50 years. The weight of evidence and agreement on causes is greatest for the short cycle and least for the long wave (for a good review, see Van Duijn, 1983; see also Schumpeter, 1939; Abramowitz, 1956; Mandel, 1975). The first is generally agreed to be an inventory cycle, the second is usually tied to investments in plant and equipment; the third is associated with construction and migration; and the long wave appears to be based on investments in long-lived infrastructure and the general buoyancy given to investment by the opening up of fundamental technologies and other major reorientations of the economy (see chapter 7).

(or between technological stasis and decline). A more important lesson is that Schumpeter failed to recognize that technology is not just a way out of crisis but also a way into it, as was foreseen by Marx. Marx (1894) tried to probe the potential contradictions between the growth of output and the production of value. He claimed that technological change is both the prime motive force of capitalist expansion *and* the principal cause of a falling rate of profit, owing to an overly rapid increase in the composition of capital (i.e., the ratio of non-labor to labor inputs). Such a shift causes the profit rate to fall because an expanding stock of capital must be rewarded from a slower-growing quantity of surplus value generated by living labor.

Marx's theory of growth and crisis has foundered for decades on the question of whether the composition of capital actually outruns the rate of surplus value. The logical force of the argument appears to expire in a series of opposing tendencies (Fine and Harris, 1979; Harris, 1983). Marx's theory of crisis runs aground because, ironically, it annuls the very developmental process which so concerned him. The tendency for the rate of profit to fall operates only within a given technological and organizational framework of the forces of production. That is, technological changes in the production process are applied to a set group of industrial sectors. With this assumption, it is only logical to conclude that the rate of profit falls as the composition of capital rises in those sectors. But technological change is only partly about the perfection of production processes in existing sectors, and for defined products. As Schumpeter was well aware, technological change also means the episodic rise of whole new types of product output, new sectors, and new labor processes. Many of these innovations effectively reduce the aggregate value composition of capital in the economy by introducing new, relatively labor-intensive activities for social labor. Moreover, neither Marx nor Schumpeter was sufficiently attuned to the possibility of technological change of a sort that lowers *both* the capital and the labor requirements of production, reducing labor time without necessarily lowering the potential rate of profit. In short, technological change really does offer the possibilities for galvanizing capitalist energies and for mobilizing capitalist economies, if other social and distributional conditions for growth are also met. It does not inexorably drive down profit rates.

Nonetheless, periodic declines in sectoral and aggregate profit rates do occur and cyclical crises are a normal part of capitalist growth (Dumenil et al., 1987; Webber, 1988). To understand growth and crisis simultaneously, we must look at the way technology and investment come together under the aegis of strong competition and capital accumulation. The real force of Marx's argument lies in the depiction of capitalists so driven to invest and adopt technologies as fast as possible in the struggle for competitive advantage that they hurl themselves collectively off a sustainable growth path (Pasinetti, 1981). Market signals and individually rational behavior are not enough to keep capitalists from collectively following the road to ruin (Shaikh, 1982). Marx develops a convincing case for how the accumulation process generates overinvestment (Devine, 1980; Harvey, 1982). As an investment boom proceeds, so do the dangers of overinvestment: too much fixed capital (overcapacity), excess inventories (both outputs and input overstocks), too much debt on firms and throughout the system. Marx calls this situation "overaccumulation:" too much capital in the system relative to the prospects for earning sufficient profits (i.e., quantity of surplus value). Profits fall, investment slackens, and the economy turns down. All that matters is that the rate of investment outruns the rate of value expansion through technological change.[21]

In the beginning of an investment boom, technical innovations will be picked up because of the business revival (tales of unused inventions lying fallow until economic conditions are auspicious are legion). As the upswing continues, productivity rises as new capacity with the latest technology is brought on line, while sales expand as new product uses are discovered and prices drop because of improved production techniques. A high rate of innovation is generally good for growth, as Schumpeter indicates. If the rate of innovation slows, the boom will begin to stall. Worse, investment will overrun profitable outlets, driving down the rate of profit. Competition at this point typically shifts to the search for cheaper factor inputs, especially labor (Rothwell and Zegveld, 1985). If a number of sectors – particularly those which may have recently undergone rapid growth in output, employment and productivity – go into decline, it drags down the whole economy.[22] On the other hand, rapid technological change alone cannot prevent crisis. Rapid innovation can attract excessive quantities of capital from eager investors. Thus even dynamic sectors such as semiconductors periodically overrun their growth potential, resulting in overcapacity and deadly competition. If the situation cannot be

[21] Most models of business cycles consider either investment (capital-stock adjustment) or technical innovation, but not the two together. Simple cycles of overproduction/overinvestment are easy to generate from models of lagged capital-stock (inventory or fixed capital) adjustment – even 50-year long waves. For a useful discussion, which nonetheless perpetuates the dual approach, see Van Duijn (1983).

[22] Industries follow their own cycles to some degree owing to their different technological potentials. At the same time, there are significant interconnections of sectoral growth rhythms because of (1) complementarity: the rise and decline of one industry affects its suppliers; (2) substitution: product development in one sector can undermine markets for others; (3) income effects: final markets of one sector depend on income generation from other sectors; (4) input–output relations: if industries' growth rates are such that they overproduce or underproduce each others' needs there will be a crisis of disproportionality (Marx, 1893; Mandel, 1975). Pasinetti (1981) drives this home by showing that every possible equilibrium growth path is immediately upset by the uneven development of technological change and consumption patterns across sectors.

corrected, the economy goes into an open crisis. For both Marx and Schumpeter crisis helps restore the conditions for growth. Marx emphasizes the way devaluation of excess capital serves as a balance wheel for the accumulation process (Harvey, 1982). Capital is written off by closing plants, scrapping equipment, selling goods at bargain prices, writing off bad debts, bankruptcy proceedings, and so forth. This disaccumulation eventually improves the prospects for profits and new investment (but see Keynes, 1936).

Growth swings are integral to the way industry develops. Massey (1979) refers to this temporal rhythm as "rounds of investment" and stresses the way the industrial landscape is shaped and reshaped over time as industry restructures from round to round. We call the erratic tracks of growth "industry development paths," because the word path suggests the sinuous, unsteady and idiosyncratic course of industrial expansion through time (and space). Restructuring is a necessary part of the rhythms of industrial growth, at both sectoral and aggregate levels; it follows from technological and other innovations that occur with expansion, as well as from the pressures of contractions. Restructuring is, in the end, just another term for the inconstancy of capitalism. Restructuring means, moreover, permanent economic disruption, as the older ways of producing die along with the birth of the new and the urgency of the modern. It is well to remember the competitors, from the handloom weavers of England to Gulf Oil, who fell from the race and were trampled beneath the wheels of progress, to recall the products and machines that became obsolescent, from cable cars to longhorn cattle, to observe the disinvestment that rolls up the rug behind the advancing front of industry, dismantling once-mighty factories from Manchester to Lowell. It is especially important to think of the human cost of uncompetitiveness, obsolescence, deskilling, and plant closures (Berman, 1982). All are swept away by the gale of creative destruction that is capitalist development. The common terms we use, such as "restructuring" and "the rhythms of growth" are ultimately insufficient, for they fail to convey any sense of the violence of economic history.

3 How Industries Produce Regions

3.0 The spatial dynamics of industry growth

This chapter concerns the ways in which industries produce regions as sites of economic activity. As we apply the growth theory set forth in the preceding chapter to the spatial dynamics of industrialization, it is again necessary to move from a theoretical universe revolving around market exchange, price signals and equilibrium to one centered on production, growth, and disequilibrium. This chapter presents a **theory of geographical industrialization**, or the microeconomic foundations of locational analysis at the industry level. We shall demonstrate that the basic patterns of industry location and regional growth can be produced by processes endogenous to capitalist industrialization, rather than by the exogenous placement of resources and consumers. Industrial location patterns are created through the process of growth rather than through a process of efficient allocation of plants across a static economic landscape. That is, industries produce economic space rather than being hostage to the pre-existing spatial distribution of supplies and buyers.

The discussion is organized around four basic locational patterns of industries: localization, clustering, dispersal, and shifting centers. Figure 3.1 schematically depicts each of these geographical dimensions of the industrialization process. We begin with the way fast-growing industries create their own locational conditions and thereby secure a measure of locational freedom from the past. Next, we consider the way dynamic economies of production lead to spatial concentrations of industry, and how growth propels the dispersal of plants from established industry centers. Then we consider the periodic convulsions that lead to long-run shifts in the centers of industrial activity; and in a closing section examine the sources of regional factor supplies.

3.1 Accelerated growth and locational windows

The dynamic process of industrialization, driven by technological innovation, organizational change, and labor intensification, and instituted

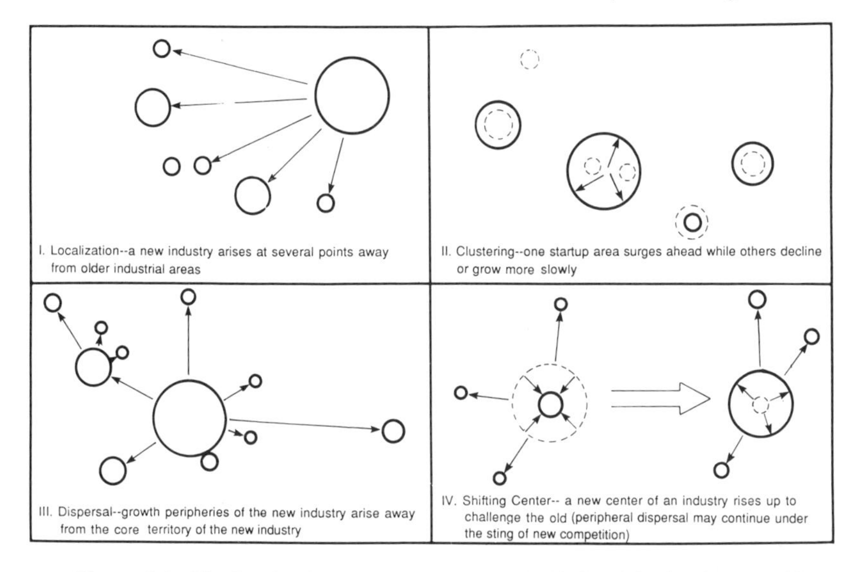

Figure 3.1 *The four basic patterns of geographical industrialization (schematic)*

by capitalists under conditions of competitive disequilibrium, enables industries to create their own geography. Contrary to Weberian location theory, industries are capable of generating their own conditions of growth in place, by making factors of production come to them or causing factor supplies to come into being where they did not exist before.[1] Capitalism is capable of escaping from the past to create new localizations of industry.

Industrial history is replete with cases of industries springing up in unexpected places, outside existing major industrial regions and urban centers. Henry Ford founded his firm in Detroit, a minor city, and made it the center of US (and world) automobile production in the 1910s by virtue of his Model T design and assembly line methods; Alfred Sloan further cemented Detroit's position with his marketing and organizational innovations at General Motors in the 1920s (Hounshell, 1984; Chandler, 1962). Cyrus McCormack moved from Virginia (then center of the wheat belt) to frontier Chicago in the 1840s to build his new reapers, while John Deere, a blacksmith from Grand Detour, Illinois, went no further than Moline to build better plows, and northern Illinois became the US (and world) focus of agricultural equipment production (Pudup, 1987). Los Angeles, known chiefly for orange groves and health spas, became a center of aircraft manufacture by World War I and moved to the top of the heap by the 1930s (Cunningham, 1951). Santa Clara County, Phoenix and

[1] We consider the question of markets for output, or final demand creation, in a later chapter.

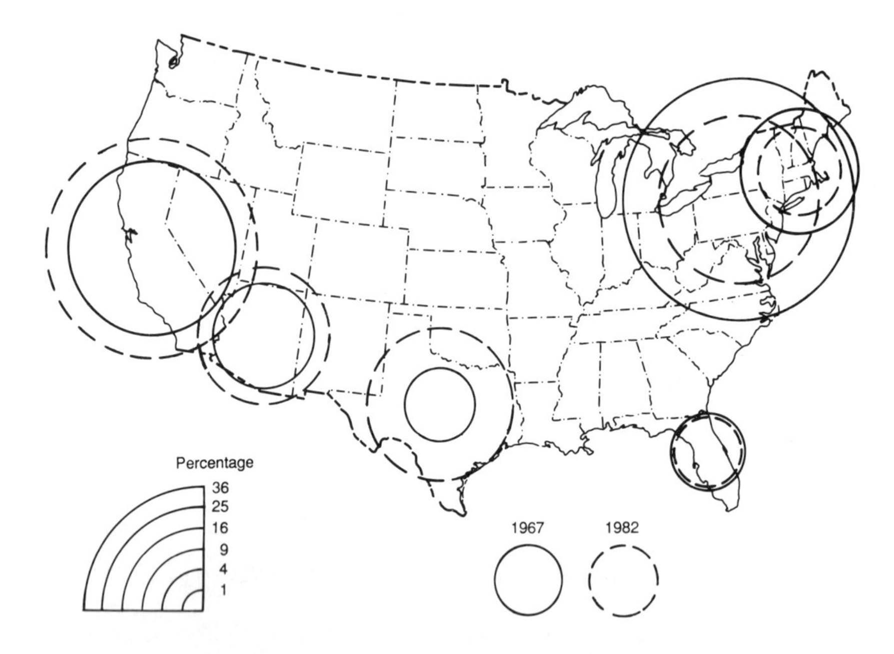

Figure 3.2 *Westward shift in the semiconductor industry in the United States, 1967-1982* Despite the large in-house production of semiconductors by the giant firms IBM and AT&T in the East, western centers have developed more rapidly since being founded by independent capitalists in the 1950s and 1960s and especially since the invention of the integrated circuit. Locational concentrations are measured as state percentages of national employment in SIC 3674 ("semiconductors and related devices"). (Data from US Bureau of the Census, *County Business Patterns*)

Dallas, all relatively obscure sites, became early homes of the new semiconductor industry in the 1950s and have prospered mightily since (see figure 3.2). New England surprised old England with its sailing ships in the nineteenth century, and West Coast ship builders did the same to East Coast shipyards during both World Wars (Wollenberg, 1988). In Japan and Korea, supertankers and video recorders are examples of new industries appearing altogether outside the dominant industrial complexes of the Atlantic economy after World War II.

Innovative firms in new or revivified industrial sectors are thus more likely to invade new or unlikely territories than might be expected from neoclassical economics and its geographic offshoots. Neoclassical models give us images of highly-constrained spatial behavior, bounded by market clearing, perfect competition and profit equalization. The locational calculus forces one to find an optimum site within the existing bounds of market areas, transport systems and resource supplies, with marginal factor substitution in response to relative transport and land costs (Isard,

1956; Smith, 1981). In contrast, the metaphor of "leapfrogging" captures the potential for new firms to lead the way into new regions, or into old, seemingly abandoned, industrial cores. Leapfrogging is the spatial correlate of capitalism's dynamic capacity to effect the radical reallocation of social resources into fast-growing, high-profit industrial sectors.

Leapfrogging presents a paradox, however: how can an industry operate and expand in areas where the costs of production are likely to be high – that is, where labor supplies are apt to be poor or inappropriate, linkages to relevant suppliers and buyers spotty, local markets weak, infrastructure poorly developed, and so forth. It is decidedly not the case that peripheral locations are cheaper sites for production, because the same conditions of underdevelopment that make labor or land inexpensive generally make them less productive in use. This dilemma of false cheapness plagues backward regions from Georgia to Ghana. If real costs were really lower in the Third World, all industrial production would have long since moved abroad. The difficulty lies in explaining why it has not, as Myrdal (1957) understood in coming up with his cumulative causation model.[2] This theoretical void has been filled by *ad hoc* lists of local attributes in new industrial centers, such as the claim that high-technology industries go to places with high "quality of life" indices (Hall et al., 1987).

We offer an explanation of the relative locational freedom of fast-rising industries. To begin, let us adopt some terminology to permit translation between conventional location theory and our geographical growth theory. On the one hand, every plant or industry has certain "locational specifications," comprised of the kinds of input–output relations that are the focus of conventional approaches. These include, first of all, the Weberian trilogy of factors: labor, natural resources, and consumers. These are non-ubiquitous and spatially-differentiated needs that vary in availability and cost at different sites. Labor stands out as the key factor, but in certain cases natural resources or special pockets of demand can be decisive. Locational specification becomes a more general and more difficult problem, however, when it is recognized that most commodity inputs and outputs pass between firms in an extensive division of labor within and between industries, relations that are usually called "linkages." Every distinct unit of the industrial system – plant, firm, or industry – has certain input–output conditions to meet, and these may be more or less well provided for at varying locations.

The term "locational capabilities," on the other hand, refers to the capacity of a plant, firm or industry to secure what it needs – labor, suppliers, buyers – at a given location. In a static context, this means

[2] Even when there is no leapfrogging, as with electronics production in the ruins of nineteenth-century industrial Massachusetts, the revival cannot be explained in terms of cheap labor due to deindustrialization (e.g., Harrison, 1984), because neither labor skills nor habits carried over from the previous era are appropriate to the new industrialization process.

ability-to-pay for inputs and ability to secure market share, both of which depend on production costs, rate of labor exploitation, and product quality; this sort of capability amounts to the same thing as competitive standing in final and factor markets, in the weak sense of competition. The meaning of locational capabilities becomes more complex in a dynamic context. Here it refers to the creative powers made possible by technological innovation, organizational advance, labor rationalization and skills development, and rate of investment.

Fast-growing industries achieve locational freedom by locational specifications and dynamic locational capabilities. To begin with, fast-rising industries enjoy enhanced locational capabilities due to above-normal profits, which allow them to attract the resources and labor that cannot be created at the new site, even if this requires paying a premium to induce labor migration or secure stable contracts with distant suppliers. As important, however, is the way a dynamic sector generates its own inputs over time rather than simply competing for a stable quantity of goods and labor-power. In early phases of development (or renewal) specific materials, parts, and equipment may be so novel that they have to be produced on the spot, or in close collaboration with suppliers on a custom basis. Machine shops or software houses that can solve such problems are likely to be widely available in developed countries. For example, automobile makers in Detroit in 1900 could draw on the parts provided by the many bicycle, carriage and ship-engine makers there (even though Detroit was not the dominant center of those industries). Distant suppliers can also be sought out by innovative merchants who begin to specialize in the needs of the industry. As a result, linkage constraints need not be overwhelming in an out-of-the-way place. The same logic can also be applied to labor: needed skills are likely to be so novel that they can be acquired only through practical experience and on-the-job training – this applies equally to prime innovator-entrepreneurs, skilled technicians, and production workers. Furthermore, new and renewed industries almost always demand new work habits, adaptability to change and commitment to the industry's success throughout the workforce, which are inculcated principally through employment in rising firms. Even natural resource inputs may be "created" anew, as when the forest products industry moved back into the woodlands of the southeastern United States because they can support plantations of fast-growing pines for the expanding chipboard market. Indeed, such innovations have caused natural resources to decline sharply over time as a percentage of input costs to manufacturing (Barnett and Morse, 1963; Perloff et al., 1960).

In other words, there is ample reason to believe that leading firms in a rising industry do not face severe locational specification constraints attributable to needs for labor, resource inputs or inter-industrial linkages for manufactured goods, because innovation necessarily means solving technical problems presented by new ways of producing, organizational problems of how to secure (or produce) various inputs, and labor problems of mobilizing and training workers. In sum, growing industries may be said to enjoy – for a certain period, defined by the rates of growth in output as a whole and rate of change in product configuration and production technique – both a factor-creating and a factor-attracting power. It follows that the lead firms in such industrial sectors have substantial freedom to develop where they are, or to locate where they please. In its early stages, a new industry can settle on any of a wide variety of locations, a choice which would not be possible under conditions of perfect price competition and ordinary cost minimization. Nor must they be drawn into the vortex of existing industrial agglomerations, as cumulative causation theory predicts, because they profit from dynamic economies of production, accelerated investment flows, and labor influx that are not necessarily dependent on the activities of firms in other sectors. For the most part, innovators in a fast-growing industry can ignore the traditional locational calculus as they devote their attention to other matters. At the same time, the new or revived industry is not yet pinned down by enormous investments in big factories and industrial complexes. These moments of enhanced locational freedom may be called **windows of locational opportunity** (Scott and Storper, 1987). They play a special role in the expansive and inconstant geography of capitalist industrialization. It matters not at all whether the proximate cause of the growth spurt is a better integrated circuit or a fashionable tennis shoe; each represents a significant opportunity for competitive advantage, high profits and growth. Nor does the analysis apply only to new industries with completely new products; revitalized older industries can experience bursts of growth due to improvements in process or product, new employment relations, or other forms of restructuring.

The potential for new industrial localizations to develop at relatively unindustrialized places has limits, of course. In real space and time, resources cannot be immediately transferred, regardless of an industry's ability-to-pay. This is especially true of highly-skilled scientific and technical workers who, regardless of wage incentives, often cannot be induced to migrate. Thus, new regional growth complexes seem to be something of a "knife-edge" phenomenon: they are often near well-developed metropolitan regions or medium-sized regional cities (tellingly, the Dutch call these areas the "halfway zone"). The former case suggests Route 128 near Boston, Orange County near Los Angeles, the Santa Clara Valley near San Francisco, or Tsukuba near Tokyo; the latter, the Midi in France, Prato in Italy, Dallas or San Diego.

Why do innovating firms spring up precisely where they do? It seems this question can only be answered for each industry by uncovering the specific conditions under which its initial innovators arose: these are bound to differ between industries, historical periods, and even national context as each nation has its own cultural, technological, and institutional conditions of entrepreneurship. It seems doubtful that there exists one model that would describe the early economic histories of all modern industries. Nonetheless, there are powerful, general reasons why these firms are not bound to appear in old industrial spaces (see also chapter 7).

3.2 Returns to scale and the development of territorial growth centers

Firms producing automobiles, farm implements, semiconductors, and aircraft initially sprang up in a wide range of sites by taking advantage of the locational windows. But not all such places ultimately developed expanding industrial complexes; some moved along different developmental trajectories or remained stillborn. For example, semiconductor manufacture began in Phoenix (Motorola) and Dallas (Texas Instruments) at about the same time as Shockley Laboratories and Fairchild Semiconductor were established in Santa Clara County in the late 1950s. Aircraft production began in Wichita (Cessna), Buffalo (Curtis), Seattle (Boeing), Los Angeles (Martin, Lockheed, Douglas), as well as Baltimore and Bridgeport. Farm machinery started up in Stockton (Holt), San Leandro (Holt) and San Jose (FMC), California, as well as in the Midwest. Yet Santa Clara County, Los Angeles, and Illinois became the overwhelming centers of attraction in semiconductors, aircraft and farm machinery, respectively. Only these places developed large complexes of firms producing intermediate inputs as well as final outputs. Some secondary centers have survived, but these are usually based around a single large factory complex such as Motorola in Phoenix or Boeing in Seattle, and most of their suppliers are to be found in the primary centers (Erickson, 1974, 1975; Glasmeier, 1985; Shapira, 1986).

How is it that some places surge ahead of others? Growth can focus around almost any outpost of a new industry, but eventually certain growth centers, or territorial clusters, gain competitive advantage over more scattered producers. Over time, this advantage increases with increasing returns, enhanced technological capacity, labor in-migration, and cumulative investment. When some places get far enough ahead, the locational window closes shut. What is wanted, then, in studying Silicon Valley, Los Angeles, or Detroit is not a series of anecdotes concerning Fred Terman, Louis Mayer, Glenn Martin, or Henry Ford, but insights into the process that propelled them into the pre-eminent centers of their respective industries (Scott and Storper, 1987).

The pattern described presents another paradox: why are plants and firms drawn toward a growing industrial cluster where factor costs may be rising faster than in other locations already staked out by the industry? This makes little sense in the terms of traditional location theory: instead of seeking the path of least cost (dispersed locations), everyone is actively bidding for land, labor, and other resources in areas of industrial concentration. The reason is the compensatory advantages (economies) of amassing large quantities of labor, machinery, materials, and structures in a limited area. Such spatial agglomeration is explained in conventional location theory chiefly in terms of minimizing collective costs of access and maximizing total revenues (Weber, 1909; Smith, 1981). This will not do. Productivity and the organization of production are at the heart of agglomeration. Productivity increases brought by the division and integration of labor overwhelm the price effects of spatial concentration. We have argued that these productivity effects are the result of increasing returns over time, in chapter 2; here we must show how these increasing returns are built up in place.[3]

Industrial complexes form, in the first instance, by virtue of the productivity of rationalized and mechanized labor processes within workshops and factories, or what may be called *internal* economies. These can only be realized with the concentration and intensification of work in one place. At the same time, industrialization proceeds on the basis of expanding productivity achieved through the social division of labor, which is the basis for *external* economies. This, too, is brought to a head through geographical agglomeration of a complex of related but diverse workplaces and firms, with their proliferating linkages. Further dynamic economies are created by an environment of technical competence and rapid technical change, and by augmenting the laboring powers of human beings themselves; these proceed most rapidly in the context of a spatially concentrated and diverse set of firms and workers.

Increasing returns spread over the whole of such territorial complexes, and are manifest in "Verdoorn effects" where each new round of growth is associated with improvements in average productivity (Verdoorn, 1980). Costs of production fall steadily on the terrain of the center, which crystallizes as the dominant production region for the new industry. In contrast, places that do not grow as rapidly suffer from the relative lack of external economies and fall farther and farther behind. Only the very largest, most vertically integrated factories and firms tend to survive on their own, while specialized workshops and firms flourish in the largest production centers. From these processes, new regional growth centers come to stand out as more widely scattered producers die out, or consolidate around growth poles to maximize dynamic – and strongly external – economies.

Myrdal's model of uneven regional development captures the dynamics of cumulative growth and decline, and is grounded in a concept of agglomeration economies as are those of Hirschman (1958) and Pred

[3] The several dimensions of agglomeration economies – technology, organization and labor markets – are considered in following chapters; for a fuller comparison with neoclassical theory, see 5.3.2.

(1966). But none of these writers enters into the realm of production deeply enough to extract the secrets of the dynamic economies of territorial industrialization. We shall try to do so, using the two-sided model of locational specifications and locational capabilities to sketch two cases: the internal economies of large integrated workplaces and the external economies of vertically-disintegrated production complexes.

The workplace as a geographical cluster The productivity of labor in a growing industry may be increased by several means: intensification of work; cooperation among workers; division of labor processes into increasingly specialized tasks; and application of better tools and machinery (Marx, 1867). To this classic list may be added improvements in product design, use of better materials, savings on inputs and reuse of wastes, and simple scale economies in physical infrastructure (Marx, 1895). Most of these advances have occurred within the confines of specific workplaces: workshops, factories, railyards, and the like. Large workplaces permit direct interaction among many workers, establishment of a detailed division of labor between tasks, the application of large and expensive machines, and the concentration of huge quantities of materials. Looked at this way, the factory or other large work site is a form of industrial clustering, for it brings together what is otherwise done in dispersed artisanal workshops and households. The possibilities for reorganization and rationalization afforded by this spatial massing of activity make possible dramatic improvements in productivity, or internal economies. Increased productivity can, in turn, stimulate demand by allowing prices to fall. As productivity grows, larger market areas can be served from a single workplace, and industry localization increases accordingly.

It would be a mistake to treat workplaces as mere points on the map, because they can include dozens of buildings, yards, canals, roads, or docks, extending to scores of acres. Furthermore, factories and other large workplaces, such as construction sites or airports, generate complementary processes of spatial aggregation by drawing into their orbits many smaller suppliers of materials, parts, machinery or business services. Typical examples are the IBM plant in Montpellier, France, where suppliers were virtually ordered to relocate nearby, and the Hughes Aircraft plant in northern Orange County, California (Bakis, 1977; Scott and Angel, 1987). Thus, most factories create larger complexes by the force of their own productivity and size. In the same fashion, big factories draw labor to themselves, even in peripheral locations such as Pullman, Illinois and Gary, Indiana built well outside Chicago at the turn of the century (Buder, 1967; Pratt, 1911). Factory placements often trigger the development of more extensive disintegrated production complexes, but can also remain more or less isolated, as in the case of the Harris missile electronics plant at Cape Canaveral, Florida (Glasmeier, 1985). Despite the overwhelming attention given to factories in accounts of the history of capitalism (e.g., Mantoux, 1961), industrial clustering has proceeded by a second major route, the disintegrated production complex.

The vertically-disintegrated production complex The growing internal division of labor and productive powers within each factory or firm is frequently paralleled by an expanding social division of labor between production units. Dynamic economies in production may be realized externally in the form of a production complex of plants and firms linked together by market (and quasi-market) transactions. In the case of the large factory, production organization is said to be vertically *integrated*; in the industrial complex, it is vertically *disintegrated*.[4] Integration occurs when the range of tasks performed in an input–output matrix are such that the division of labor, even when very elaborate, lends itself to internalization within a single factory or under one managerial hierarchy within a single firm. Under certain conditions, however, discrete labor processes may separate into specialized workshops and firms (see chapter 5).

The reasons for disintegration of the production process and reliance on external economies are many and powerful, and often typify the situation of a rapidly growing industry. First, many labor processes resist integration into unified machine systems (i.e., they are not technically indivisible) (Williamson, 1975). Second, the optimal scales of operation for different subprocesses may not match, leading to excess capacity and inefficient utilization of some (Scott, 1983). Third, inputs of specialized goods and services may have optimal scales of production which can only be attained if they are spread over a number of downstream firms (Holmes, 1986). Fourth, where output markets are unpredictable or unstable, producers may disintegrate in order to avoid transmitting uncertainty through the vertical structure of the firm (Berger and Piore, 1981). Fifth, final output may require inputs that can be manufactured most efficiently by firms that devote specialized, managerial attention or knowledge to the task (Vennin and de Banville, 1975). These conditions are likely to apply very frequently to industries in which product configurations are still pliable and where markets are rapidly changing.

This is the process of increasing roundaboutness of production alluded to in chapter 2. Figure 3.3 illustrates the process of deepening social division of labor and increasing roundaboutness of production in a growing industry. To begin with, there are only two producers, one providing an input to the other. As the industry develops, not only does the number of workplaces and firms increase, but linkages become more complex as firms develop multiple trading partners to spread risk and uncertainty,

[4] The term "vertical (dis)integration" is a simplification of more interwoven patterns of input–output (and other) relations. For a formal analysis see Pasinetti (1981).

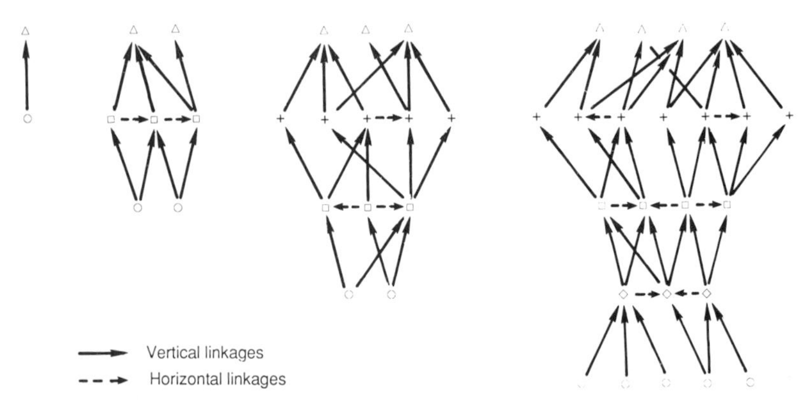

Figure 3.3 *Vertical and horizontal expansion of the division of labor in a hypothetical industry* This schematic diagram illustrates the evolution of a production system from two establishments in two sectors to twenty-three establishments in five sectors, and the resulting increase in the density of linkages among units. (After a draft by Allen Scott)

and as new layers in the input–output system develop, composed of increasingly specialized workplaces. Horizontal linkages also develop where firms attempt to offset variability in demand by purchasing from other enterprises that portion of output which cannot be predicted or stabilized over time.

In growing industries, roundaboutness generates dense and complicated transactional relationships between firms. The greater the transactional costs, the greater the likelihood that producers will agglomerate in order to reduce them, because spatial proximity is still a fundamental way to bring people and firms together, to share knowledge and to solve problems. Disintegrated production complexes thus appear on the landscape of capitalist production with some frequency; they are the geographical means by which industries most easily realize increasing returns in the externalized form of a social division of labor.

Highly-agglomerated, vertically-disintegrated industries are found frequently in inner cities around the globe, as with clock-making in nineteenth-century London (see figure 3.4), or clothing, jewelry and toys in downtown Los Angeles and motion pictures in Hollywood today. But they are also to be found at a wider geographical scale in the form of regional industrial clusters, as in the clothing and knitwear industry in Prato outside of Florence, the machinery industries outside of Bologna in Modena, or the nineteenth century hosiery industry of the East Midlands of England.

As the agglomerated territorial complex grows, spinning off new activities and firms, it, too, creates its own locational specifications: a network of intermediate buyers and sellers, users and suppliers of each of these products. Linkages, and the spatial relations necessary to their

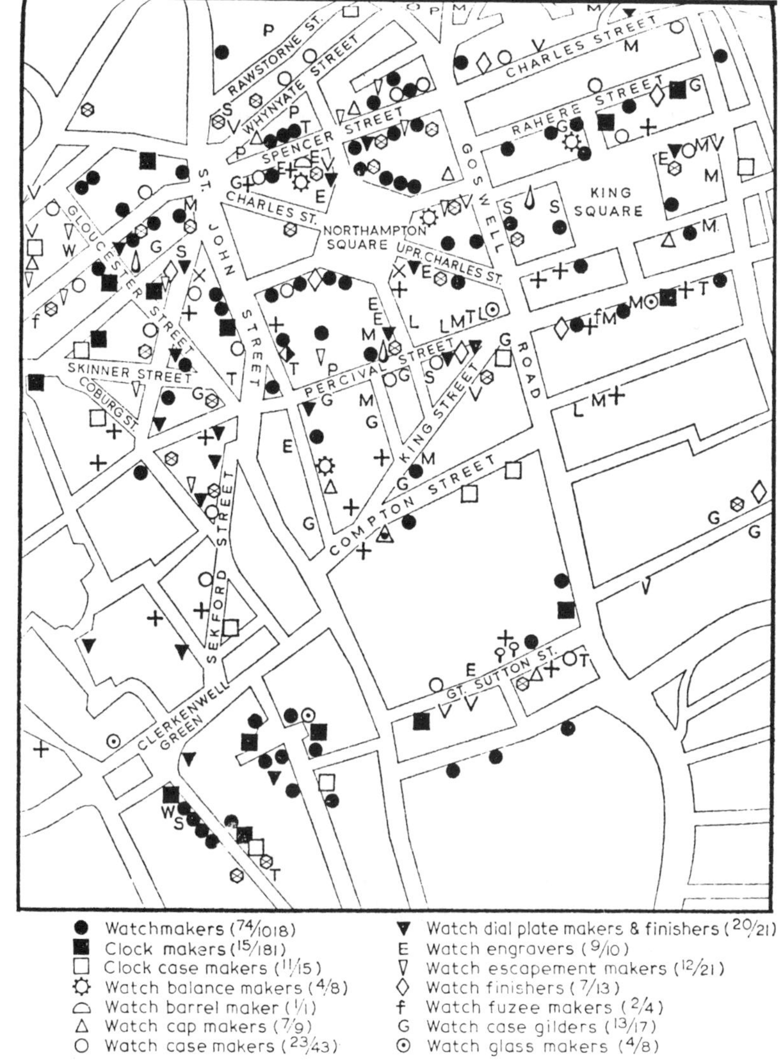

Figure 3.4 *Agglomeration in the London watch- and clock-making quarter of Clerkenwell, 1861* The map shows the clustering of diverse watch- and clock-related crafts, particularly parts suppliers, in a single industrial district of a large metropolis. Figures in brackets indicate the number of establishments in the map area relative to those active in each trade for all of London. (From Hall, 1964)

fulfillment, thus can develop simultaneously; location is not something that follows once input–output coefficients are established, as in Weberian models. On a larger stage, the growing cluster attracts sellers, merchant intermediaries and labor from afar even more vigorously than the large factory, while its output markets expand through product improvement and cost reduction.[5] Labor supplies are also largely created within these budding industrial complexes. For semi-skilled workers, the local labor market functions like a giant training mechanism and hiring hall, even in times of high labor turnover and changing labor processes (Shapira, 1986). Chains of migration bring in additional workers and even periodic oversupply, helping the industry keep down wage costs by checking the tendency for labor to bid up wages (Thomas, 1973). For skilled engineering and technical labor, the shifting social division of labor and continual innovation constitute an "opportunity grid" to develop skills in practice, as skilled workers engage in product research, innovation, design, and manufacture.

In any new territorial growth center, moreover, processes of habituation to the peculiar rhythms and imperatives of work in the local area are facilitated by the transmission of norms, culture, and consciousness formed around the experience of work. The community is the repository of social practices and attitudes that facilitate socialization of the laborforce and provide important cues to workers about legitimate expectations and habits. Socialization is further enhanced where, as is typically the case, specialized educational institutions and training establishments develop nearby, as with the electrical engineering programs at Stanford and Berkeley in the Silicon Valley area. In short, there are very strong reasons for specialized complexes of industrial activity to be able to produce their laborforces *in situ*, in effect drawing into their orbit their primary locational specification, making the laborforce an historical creature of industrial activity, and not an independent locational factor.

The history of capitalist industrialization is replete with spatially-dense industrial production complexes beginning with cotton textiles in Lancashire, cutlery in Sheffield, and silk in Lyon, and continuing to the present-day wave of high-technology industrial districts (Sabel and Zeitlin, 1985; Scott, 1988a). The ability of these industries to take advantage of windows of locational opportunity by producing their own locational specifications, means that they often effectively expand the spatial range of advanced industrialization. Once these complexes are established, moreover, they are bound rather firmly to the spot by the immense fixed capital they lay down, the established community of labor and the web of linkages every firm and workplace has to other units in the social division of labor. Thus, what may have begun in essence as a short-term strategic choice, or even an accident, that permitted one complex to get ahead of all the others may ultimately have durable effects on the shape of entire space economies, as the window of locational opportunity shuts with the

[5] Smith's (1776) classic dictum that "the division of labor is limited by the market" – taken up by modern neoclassical theorists (Stigler, 1950) – is therefore wrong. The market – including both final and intermediate demand – is just as much limited by the division of labor and its effect on productivity and product innovation.

generation of massive place-bound external and internal economies of scale and processes of social reproduction.

3.3 Dispersal of established industries

All the attention given to deindustrialization of core regions in recent years, and the industrialization of various Sunbelts, new development zones, and newly-industrializing countries (NICs), has created the impression that industrial activity has transcended spatial concentration to become widely dispersed. Not so. Most industrial activity remains concentrated in a few regions of each country and in a few countries in the world, and regions still tend to have rather specialized industrial bases. In addition, the industrialization of Sunbelts and Third World areas is based largely on different industries from those in the old manufacturing belts. Generalized dispersal – industrialization that does not take a territorially localized and agglomerated form – is a strictly secondary phenomenon.

Nonetheless, dispersal of industries from their dominant growth centers is common. This may take the form of entire firms moving to new locales, but characteristically top management, key research and service activities, and some production remain in the historic center while new manufacturing, marketing, and divisional headquarters spring up in distant locations. This pattern has been explained in several ways, all basically variants on two models: the neoclassical and the product cycle. They yield certain insights, but ultimately fail to situate dispersal within the framework of capitalist growth and industrialization, as they give excessive weight to exogenous location factors or simple forms of technical change. As a result, they do not explain the main patterns of industry decentralization.

The standard deagglomeration argument focuses on prices as the drivers of business behavior. In this view, as relative prices tilt against the center and in favor of the periphery, industry slides down to a cost-minimizing location. In one variant of this approach, the core region (e.g., the northeastern United States) has a historical head start on industrialization, but loses this over time as industrial activity spreads out to the backward regions (e.g., the Sunbelt), until development is evenly distributed over the country (Borts and Stein, 1964). Unfortunately, the evidence of long-term evening-out of regional incomes is meager (Perry and Watkins, 1977; Persky, 1978). In another variant, industry initially congregated in core cities and regions to avoid high transport costs, but has dispersed rapidly in the twentieth century thanks to the flexibility and speed of truck traffic (Chinitz, 1960; Moses and Williamson, 1967). But suburban and inter-regional dispersal of industries antedates the truck, and metropolitan aggregations have not disappeared with modern transport and communications, they have only spread out farther (Walker, 1977, 1981; Scott, 1988a).

A third variant argues that as industrial complexes grow, costs of production tend to rise owing to "diseconomies of agglomeration." For example, as semiconductor and computer firms crowded into the Santa Clara Valley, land and housing prices, commuting distances and times all increased, labor turnover rose, and most recently, growth controls and new environmental regulations have been imposed (Saxenian, 1984; Siegal and Markoff, 1985). Such costs have frequently been invoked to explain why electronics companies move out of the Valley, but in fact, the main push of semiconductor assembly to Southeast Asia occurred in the late 1960s and early 1970s, well before Santa Clara County experienced severe congestion and housing inflation. A fourth variant emphasizes labor cost and militancy as the chief cause of dispersal (Gordon, 1977; Peet, 1984). Yet here again, one finds rapid dispersal taking place from Silicon Valley and other industry growth centers well before unionization or other signs of widespread labor militancy. A final variant projects the development of transport and communication networks and the multinational corporation onto the international stage as factors increasing the locational capability of capital, while labor is locked in brutal competition between advanced and underdeveloped nations; this is sometimes called, the "new international division of labor" (Froebel et al., 1977; Peet, 1987). We share this concern for the plight of workers, but these left models still rely on factor-induced behavior over growth-centered locational dynamics.[6] There are two principal flaws in this sort of reasoning. First, productivity gains usually continue to outstrip diseconomies of the agglomeration – which is why industrial agglomerations are possible in the first place (Kaldor, 1970). We shall say more about this in chapter 5. Second, industrial deagglomeration is virtually always associated with changes in production techniques and reorganization of production processes, such that the mix of factors required from the "locational environment" changes. It is misleading to claim that firms use locational change as a spatial means of effecting factor substitution; rather, both relocation and changes in the mix of factors used are consequent upon technological change and reorganization of the production system. This is discussed further in chapter 4.

Some of the more sophisticated versions of traditional location theory have attempted to incorporate processes of technological change. Following neoclassical models of "induced" technological change, the production function is adjusted in response to shifting factor prices, where technology permits the substitution of cheaper factors for more expensive ones (in terms of marginal unit costs). In the short run, this involves moves along

[6] It should be noted that all these views of dispersal are used by capitalists to justify policies designed to keep costs, especially labor costs, to a minimum. Such arguments for maintaining a "good business climate" form the conventional wisdom of most economic development planners from Arizona to the Philippines (Logan and Molotch, 1986).

the production function; in the long run, moves between production functions. Location models add a spatial dimension factor, where technological change responds to supply conditions at a range of different sites. Relative factor prices and quantities are the central force in economic development, as in all neoclassical theory, but now the spatial differentiation of factor supplies is a critical motivation for adjustments toward equilibrium (Isard, 1956; Borts and Stein, 1964; Dollar, 1986).

Even the most sophisticated neoclassical view of locational behavior is still essentially incorrect, however. The role of spatial factor prices is subject to the same forces that circumscribe the role of factor prices generally in technological change, as detailed in chapter 2. Guided by strong competition and the search for advantage, and subject to increasing returns, technological change does not necessarily economize on any particular factor but reduces production costs generally or creates new revenues through product innovation. Moreover, technological search under conditions of increasing returns does not necessarily lead toward optimal and predictable choices, but instead is inflexible, irreversible, and non-ergodic in nature. Since expanded reproduction does not require a reduction in use of any particular factor, outcomes are unpredictable and often appear perverse with respect to relative factor prices, including regional price differences (Nelson and Winter, 1982).

If our logic is correct, it is not surprising that induced-innovation models of spatial equilibrium do not jibe with the facts. As Leontief (1956) observed, regional patterns of production (factor intensities) are frequently contrary to relative factor prices; that is, capital-rich countries like the United States may export principally labor-intensive products, and modern capital-intensive factories may frequently be found in the cheap labor peripheries of the world economy.[7] The converse – that capital-intensive technological development is a function of higher labor costs – has not been proven either, even in the well-known debate over American technology in the nineteenth century (Habakkuk, 1962; Temin, 1966; David, 1975). In short, there is no compelling evidence for a systematic relation of factor costs to factor intensity in production methods among different regions or nations (Persky, 1978). Factor prices are strictly subjacent in a theoretical account of the dynamics of industrialization of regional peripheries.

Product cycle theory tries to respond to Leontief's paradox of the possible inverse relation of factor prices to location: labor-intensive activities that

[7] Of course, capitalists often use less mechanized production techniques in Third World plants to take advantage of low-cost labor supplies, as with semiconductor assembly operations in Southeast Asia (Dosi, 1984; Scott, 1987). But direct spatial-technical substitution usually amounts to using inferior techniques, in the form of second-hand equipment, not the invention of better capital-saving techniques (Clarke, 1985).

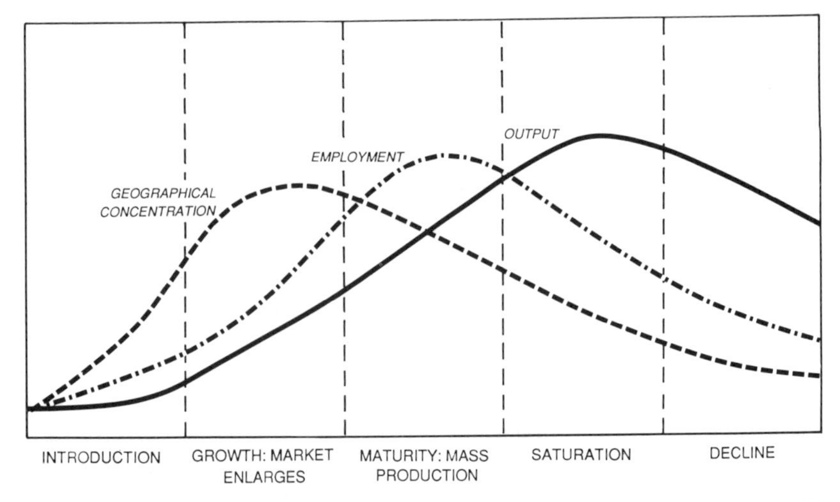

Figure 3.5 *Geographical dispersal of an industry according to the product cycle model (production maturation version)* A new product is introduced and a new industry arises. Output of the industry expands through a period of mass production until the market is saturated; finally demand declines as the product is replaced by newer ones. Employment grows along with output until advances in labor productivity drive it down. The industry initially clusters in its place of origin, but decentralizes thereafter as capital-intensive branch plants seek lower-cost locations.

depend on skill for product and process innovation are found in developed cores; capital-intensive branch plants seek out underdeveloped peripheries because factor costs are cheaper there (see figure 3.5 and section 4.4). The logic is that industries standardize and automate their production processes as they mature, which results in a deskilling of jobs (Hoover, 1948). The necessary condition for such dispersal is the ability to break dense linkages by internalizing larger portions of production systems within large factories and giant corporations and standardizing external transactions. The product cycle model appears to fit the behavior of several major industries in the postwar period, such as radios dispersing from New York City to Midwest branch plants (Vernon, 1960), or the shift of canneries out of the San Francisco Bay Area to the Central Valley of California (Cardellino, 1982). Product cycle theory has therefore been quite popular in geography, along with a more general concern for the role of "branch plants" in regional growth in backward areas (Rees, 1979; Erickson and Leinbach, 1979; Watts, 1981).[8] Nevertheless, the product cycle does not account adequately for the most important types of industry dispersal.

Most industries spin off **growth peripheries** from their centers of activity. This term captures the link between the conditions of growth in an industry

[8] The branch plant literature usually combines factor cost differentials, organization theory and the product cycle in varying degrees.

Figure 3.6 *Development of a growth periphery: the Japanese automobile assembly and supply complex in the United States in the late 1980s* Japanese car firms, led by Honda, have moved into US territory to secure their expanding North American markets. They have induced many parts suppliers from Japan to join them, as well as buying from US firms nearby. Data assembled from several sources, including interviews, by Mair et al. (1988). (After Mair et al., 1988)

and its spatial expansion. The reasons for establishing new peripheries depend on the exact conditions of growth in any sector, which change over time. The "expansive periphery" represents an extension of the successful industry and its firms into new territory to capture new markets and to eliminate competitors who have not made the same innovations. The expanding firm may build new plants on the most advanced basis or buy out failing competitors and convert their plants. This is the original "Fordist" model of geographical industrialization: soon after its first assembly line plants were built in Detroit (not in outlying areas), Ford moved to new conquests, entering Britain with a plant in London as early as 1917. Colt and Singer made similar conquests of the British market in the nineteenth century (Hounshell, 1984). General Motors, by contrast, moved abroad through mergers, such as its takeovers of Vauxhall and Opel in Europe. In California, Detroit's auto makers followed both strategies in drying up the independent base of a once-thriving center of automobile production and innovation, and converting what was the country's second center of vehicle and parts production into a growth periphery of an industry centered on the upper Midwest.[9] Today, Japan is penetrating US territory with its own expansive wave of plant construction and acquisition in an area centered south of Detroit,

[9] Thanks to Rebecca Morales for pointing this out.

principally in Ohio (Mair et al., 1988). Such dispersal does not involve only branch plants, but often generates large secondary clusters of activity (see figure 3.6).

Factories therefore often move into new territory not as a way of disconnecting from, or dispensing with, the industry core area but to extend it into new growth peripheries. When new textile mills were built all over New England at available water power sites in the first half of the nineteenth century, or when new grain mills spread out from Minneapolis to the lower Midwest after World War I, or when additional steel mills extended west from Pittsburgh to Ohio, Indiana, and Illinois early in the century, it was not really a process of disaggregation. The net result is to extend the agglomerative field of the core production complex through long-distance inter-plant transactions. At the same time, conquest of these new markets frequently generates additional real growth in the industry's core production complex – a derived demand for inputs, components and capital goods to serve a more far-flung production system. In the automobile industry for much of this century, decentralization of branch plants was thus paradoxically associated with *increasing* dominance of the Detroit region. Both the implosion of an industry into an expanding center and the explosion outward into new regions and countries are based on the same general conditions of competitive advantage and rapid growth of an industry and its firms. This is the kind of pattern that Vernon (1966) was trying to capture in his model of the international product cycle, formulated at a time when US firms had conquered European and other foreign markets. But this dispersal did not involve new products as much as the spread of Fordist and other mass production methods in a period of overwhelming competitive advantage and rapid investment by US capitalists (Littler, 1982). The general principle, of which the product cycle theory captures only a portion, is that spatial expansion depends upon growth – regardless of whether it derives from product innovation, process change, high exploitation of labor, or industry reorganization.

Dispersal may also come as a consequence of decreasing rates of accumulation in previously expansive sectors. This kind of movement appeared at the end of the long cycle of post-war growth, and was propelled by overinvestment, heightened competition and dwindling profits. This dispersal to "cost-cutting peripheries" has been erroneously conceptualized in terms of industry life cycles and the new international division of labor. In Markusen's (1985) "profit cycle" model, a variant of the product cycle, the halcyon days of product youth and rapid industry expansion are followed by slowing rates of growth, as production capacity approaches market demand, production techniques standardize, and market price/production cost margins begin to come down; as a result, sectoral competition increases, oligopolistic practices break up, and profit rates drop toward (or below) economy-wide averages. As industry and firm superprofits disappear, factor prices become a more important dimension of competition and location; to make things worse, new competitors are likely to enter the market from alternative locations (usually foreign national bases) with lower factor prices. As a result, firms try to cut costs by building new plants at cheaper locations and with more efficient production methods. The US Southeast and Southwest are littered with hundreds of such branch plants (McLaughlin and Robock, 1949; DeVyver, 1951; Hansen, 1979, 1981). At the same time, older, less viable plants in the historic centers are likely to be closed down, further accentuating the tendency toward decentralization.

Profit rates and competitive conditions are not bound narrowly to industry life-cycles, however. The competitive-squeeze effect is the outcome of a combination of technological change and investment, as argued in chapter 2, and cost-cutting strategies are likely to accompany every major cyclic decline (Rothwell and Zegveld, 1985). The plight of the US steel or auto industries in the 1980s is not due only to the maturity of those sectors and their markets, but has been compounded by overinvestment and excess competition on a world scale. The relatively disadvantaged position of US producers, furthermore, is due more to technical backwardness and failure to invest in new methods than to high factor costs. After all, their Japanese and Korean counterparts are still expanding economically and geographically. The noticeable dispersal of personal computer assembly and components production to East Asia in the early 1980s obviously cannot be attributed to the end of this industry's life; it was due, rather, to heightened competition from new growth peripheries in a fast-rising industry that overshot product demand in a rapid cyclic boom and bust circa 1980–6.

A different cause lay behind the widespread dispersal of semiconductor assembly plants by US firms in the 1960s and 1970s (see figure 3.7). Cheap labor unquestionably attracted companies to Mexico, Southeast Asia or Morocco, but here was a fast-growing sector that ought not to have felt pinched by factor costs. Indeed, the part of the industry protected by the exceptionally high profits of IBM or AT&T (the so-called "captive" producers) never did disperse assembly operations abroad (Sayer, 1986a). But intense competition and low wages are quite compatible with high rates of growth. The open-market companies (called "merchant" producers) felt the sting of competition and the need to keep assembly costs down under the peculiar technological circumstances of rapid product evolution, high research and development expenditures and high wages to technical workers, accompanied by a collective strategic choice to put off investment in the fixed capital necessary to automate assembly. (Japanese semiconductor makers, like US captive producers, shifted to mechanical assembly sooner.) Neither relative factor costs nor profit cycles can alone explain this peculiar form of growth periphery. Nor can they account for the subsequent evolution of certain Southeast

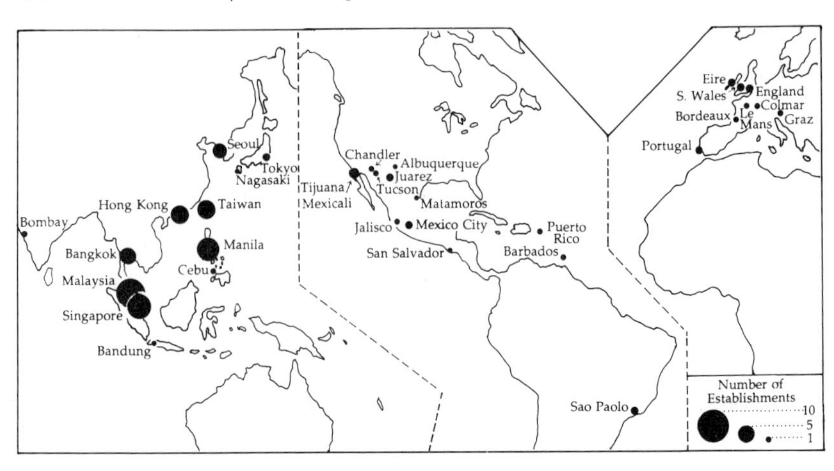

Figure 3.7 *Growth peripheries of the US semiconductor industry* Since the early 1960s, US semiconductor manufacturers have dispersed their operations widely to reduce costs and capture world markets. This map shows free-standing assembly plants only (offices, laboratories, wafer fabrication and subcontracted assembly operations are not shown). These plants have been heavily concentrated in low-cost areas such as Malaysia, the Philippines and Mexico. Even in Europe, where the market is the principal draw, they tend to be positioned in low-wage peripheries. Significantly, many of the Southeastern growth peripheries have recently been built up into fully-integrated fabrication and assembly operations to serve growing Asian markets, while no new plants have been added in Mexico, despite its continuing low wages (Scott, 1987). (After Scott and Angel, 1987)

Asian semiconductor peripheries into more full-blown centers of design, fabrication and assembly for expanding regional markets (Scott, 1987).

3.4 Restructuring and shifting centers of industry

Many facts of geographical industrialization do not fit within the tidy framework of any single cycle of development: there is no natural course of industry growth. The most dramatic paradox of industrial location is the way industries centered for decades in "optimal" spots with strong agglomeration economies, enormous sunk capital, and a stable laborforce suddenly melt away. This has occurred throughout the US rustbelt, and in northern Britain and northern France, for instance (Bluestone and Harrison, 1982; Martin and Rowthorn, 1986). Yet in most of the devastated regions nothing obvious has triggered such massive relocations. If diseconomies accumulated slowly, they apparently showed their face very quickly. Within a very few years, or even months, the economic base of a community can disappear, leaving the empty hulks of obsolete factories and decaying infrastructure.

Various theories have been put forward to explain rapid deindustrialization. Bluestone and Harrison (1982) attribute it chiefly to corporate ability to shift capital around, but this only improves site optimization given existing factor differentials. Markusen (1985) tries to explain it in terms of the breakdown of long-standing monopoly power which allowed large companies to protect their profits. Casetti (1979) borrows from catastrophe theory to argue that sudden upheavals are caused by long accretion of small changes (cf. Clark et al., 1986). The alternative theory of "industrial restructuring" argues that industries in crisis will make various strategic moves to rationalize production and revive sagging profits in the spheres of finance, organization, production, and employment, including consolidation, removal of excess capacity through plant closures, layoffs, increased automation, and work reorganization (Massey and Meegan, 1982). Restructuring theory, however, has heretofore been unduly bound to the circumstances of its discovery: the catastrophic decline of industrial Britain since 1970 (Hudson et al., 1983).

The concept of restructuring can be extended to industrial renewal and growth at the expansive frontiers of national and global capitalism. We have shown the growth prospects and locational freedom of new industries, but these industries do not come out of nowhere. They branch off from existing sectors, either through a flowering of technical possibilities (as in refining giving birth to plastics and pesticides), a shift in fashion (as with running shoes), or through branchings from older industries (as in the development of locomotives from textile machinery or missiles from aeronautics). Entirely new products are not the only means to industrial growth, as product cycle theory implies. Industries can be renewed through dramatic product modifications, revolutions in production methods, or complete reorganizations. This has been called "dematurity" of an industry – though the term seems an awkward inversion of an inapt organic metaphor (Abernathy et al., 1983). In any case, examples of renewal are easy to find: the impact of jet aircraft on air transportation after 1960, the effect of the tugboat and barge on a moribund water transport industry around 1900, or the revival of an endangered movie industry after the studio system dissolved in the 1960s. In other words, after a period of decline, growth can begin again on a new basis.[10] In short, disequilibrium growth also means unpredictable industrial shifts due to product change, production breakthroughs, crisis and restructuring.

When growth is re-ignited, the geographical implications are potentially dramatic. We are back in the position of locational freedom: in cases of significant restructuring and renewal, the locational window can open up despite the existence of a center and growth peripheries. The innovations and new firms that are the basis of the renewal will likely be found away

[10] As Gold (1964) has shown, Burns (1934) was only able to derive a tidy product cycle pattern of growth and decline by conveniently defining every point of renewal as the beginning of a new product cycle!

from the traditional center of activity, and a new center will begin to form around them and outcompete the old one. While the old center may hold on for a long time (especially if protected by national boundaries and trade barriers), the phenomenon of the **shifting center** will eventually be apparent to all. Examples are Tokyo-Nagoya out-producing Detroit in cars by the 1980s (Cusumano, 1985; Scott, 1988a), the Midwest taking New England's mantle as the center of American metalworking after 1900 (Hounshell, 1984), and the Carolinas becoming the center of US cotton textile production after eclipsing New England and Philadelphia by the 1920s (Wood, 1986) (see figure 3.8).[11]

Geographical industrialization is subject to surprising turns as industries shoot off in new directions. Long-term "industry growth paths" are inherently open-ended, irregular and subject to upheavals, reversals and new directions. Only these terms will let us come to grips with the volatile history of industries.[12] A good example is the geographic inconstancy of meatpacking in the United States over the last century, as shown in figure 3.9. Its first locus of production was the Ohio Valley, especially Cincinnati, in the mid-nineteenth century. By the Civil War, the industry had shifted west toward the Mississippi, and Chicago became the overwhelming center of slaughtering and packing for the next 50 years. Through the first half of the twentieth century, meatpacking looked like a classic case of a mature industry, developing new growth peripheries in such outlying sites as Omaha, Nebraska, Sioux City, Iowa, and East St Louis, Illinois. In the post-war era, however, the industry has again been on the move, shifting farther west into Iowa, Nebraska, Kansas, and Texas. Chicago was virtually stripped of its packers and slaughterhouses by the 1960s.[13]

[11] The classic case of textiles illustrates the dilemma faced by existing theories. The move of textiles to the US Southeast has been carefully explained in terms of the product cycle by Hekman (1980a, 1980b), but he can only fit the evidence to his model by ignoring certain contrary facts: the prior establishment of a large textile industry in Massachusetts, independent of the British industry, by virtue of the Waltham system of integrated production; the buildup of a growth center in New England on the basis of highly-integrated factory production; the rise of a second complex at Philadelphia that followed a completely different disintegrated production strategy; the rise of North Carolina as a challenger not because of decentralization from the North but through its own devices; the complete collapse of the northern centers and companies; the lack of a clear process of standardization accompanying the move South; the recent continued dispersal to the Third World; and the futility of dating any of this in terms of a "maturation" process: did maturity come to the textile industry with Jacquard looms, polyester fibre or water-driven shuttleless looms?

[12] Lloyd and Dicken (1977) speak of "break points" in the location of industries, but in the sense of comparative statics rather than in terms of continuous long-term evolution.

[13] For further discussion, see section 4.3.

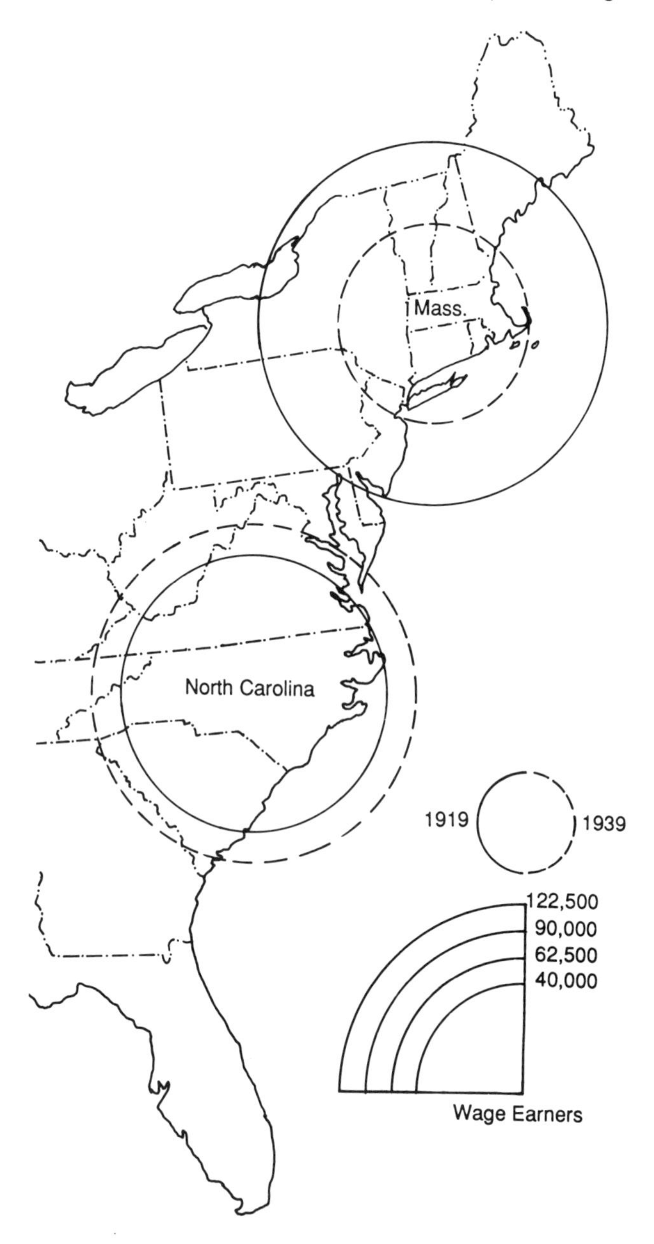

Figure 3.8 *The shift of the US cotton textile industry from Massachusetts to North Carolina, 1919-1939* The movement of textile production to the South was very rapid after the turn of the century, leaving only remnants of manufacturing in what were once the overwhelmingly dominant centers of the industry. The decline of southern New England and of Philadelphia (not shown) parallels that of Massachusetts. (Measured by wage-workers employed; based on data in Wood, 1986)

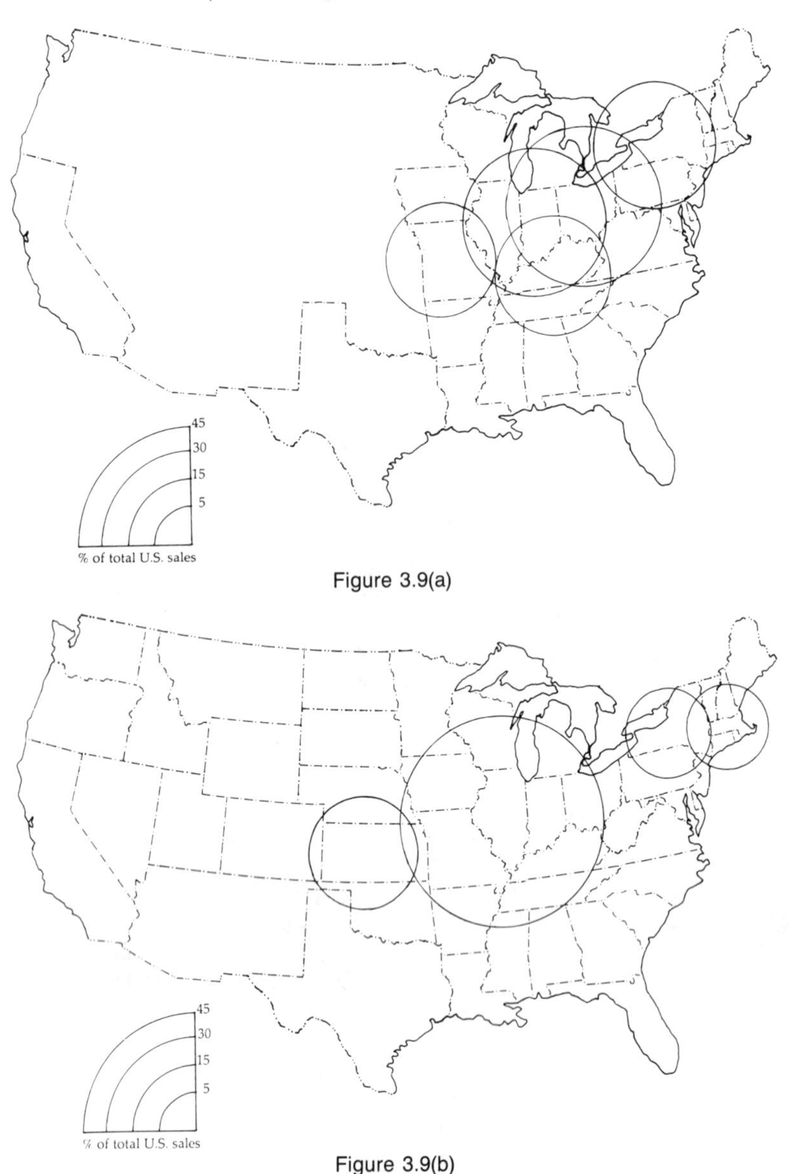

Figure 3.9(a)

Figure 3.9(b)

3.5 The production of regional resources

We have thus far argued that industry location patterns are to a large degree independent of regional factor supply conditions. We now examine the region, and how it acquires its endowment of resources and income. In the production-centered growth model proposed here, regional factor

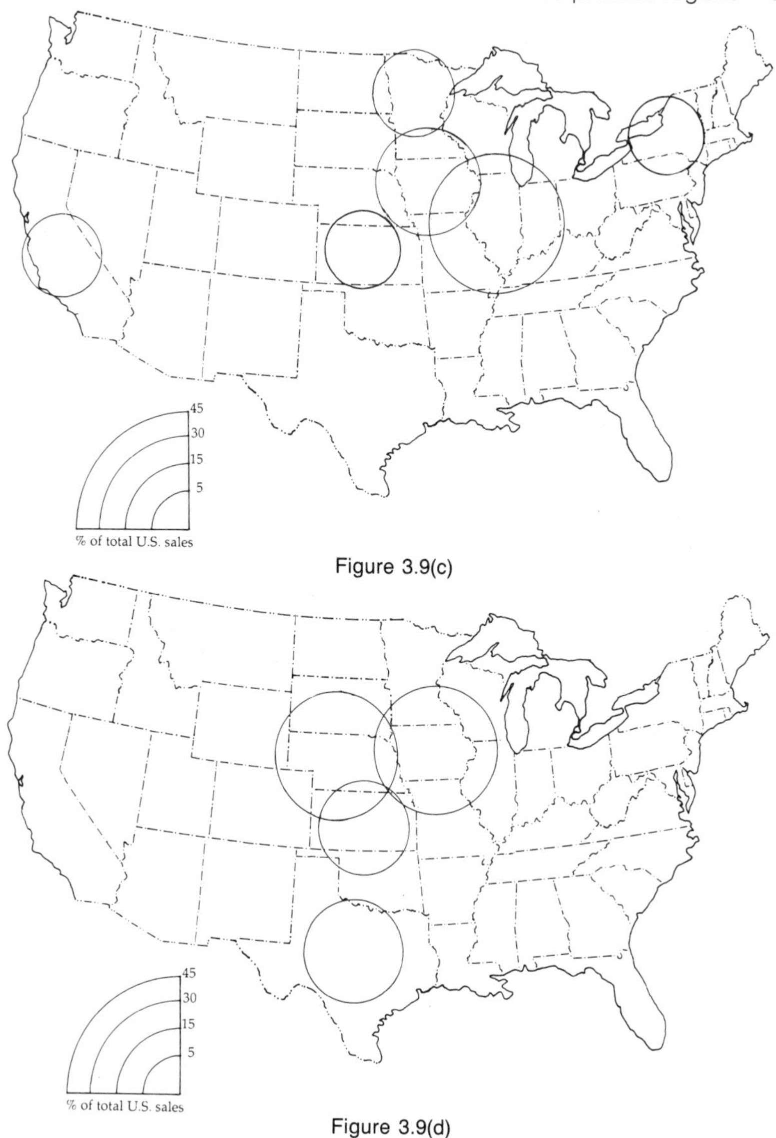

Figure 3.9(c)

Figure 3.9(d)

Figure 3.9(a–d) *Locational shifts in the meatpacking industry, 1850-1982* Slaughtering and meatpacking were concentrated in the Ohio River Valley in 1850. By 1890, Chicago (Illinois) was the overwhelming focus of the industry. By 1940, the industry had developed several growth peripheries, while the core area of Chicago had shrunk considerably. A complete recentering had taken place by 1982 to a band of states from Iowa to Texas. Figures are percent of US sales, by state; only states with over 5 percent of the industry are shown. (Data from US Bureau of the Census, *Seventh Census*, 1850, *Statistics of Manufacturing* (1859), p. 94; *Eleventh Census*, 1890, v.6:1, *Manufacturing Industries*, table 313; 1939 *Census of Manufacturing* (1940), pp. 54 - 5; 1980 *Census of Manufactures* (1982), Industry Series 20A - 7)

prices and quantities do not determine the region's mix of economic activities. This turns the neoclassical world on its head, and views regional development as the result of regional capital accumulation through the region's particular industrial history; developmental dynamics are theoretically prior to and different from the allocation mechanisms of market-centered models. That is, industries create regional resources and not the other way around. The composition of inputs in an industry and the scope of the market are the result of innovations in product, process, and organization that generate competitive advantage, dynamic economies, and high rates of accumulation. Moreover, innovating firms and burgeoning industrial clusters develop many of the materials, and much of the machinery and labor skills they require, on the spot or through close interaction of user and supplier firms. Thus firms and sectors generate their own input histories, and those of their chosen regions, at the same time. It follows that the central motor of regional development is not industry location as a response to prior resource endowments, but geographical industrialization as a process of growth and resource creation.

Factor prices and regional income differentials also tend to follow the course of industrialization, not the reverse. As argued in chapter 2, prices and profits are consequences of industry development paths, including rates of demand expansion, productivity growth, and investment. The center of gravity of long-run regional factor prices will be set by the parameters of production and investment, with factor payments responding *ex post* to variations in profit rates between firms and industries, as they are distributed in geographically uneven fashion. Continual technical change, and disequilibrating forces in general, mean there is little possibility for factor supplies to adjust sufficiently to equalize either their own rates of payment across space or inter-firm/inter-industry rates of profit (Sheppard and Barnes, 1986).

Labor, whose income share is in part politically determined, will consciously seek to adjust wages in light of observable rates of profit and growth. Even without this conscious intervention, relative permanence at growth centers and peripheries is likely to set up a cost-of-living cycle in the local economy which, when combined with the spatial division of labor among industries, reproduces wage differentials (Clark et al., 1986).[14] High-wage regions are likely to have consistently higher levels of average productivity than low-wage regions, owing to greater industrialization and accumulation of capital. Low-wage, resource-scarce

[14] Most industrializing regions embrace more than one industry, and regional competition and factor markets tend to average the various resource demands of different industries. This may influence long-term regional factor prices, if they become incorporated into a regional cost-of-living cycle (Cebula and Smith, 1981). On the other hand, local labor markets can be remarkably distinct even within dense industrial regions and metropolitan areas. Different spatial patterns of industrialization can affect regional income distribution, and thereby local and national growth experiences (see chapters 6 and 8).

regions are largely that way because they are poorly industrialized, not the other way round. Comparative advantage and disadvantage are both historical products of geographical industrialization.

While it is not necessary to reject altogether the conventional market categories – supply, demand, and price adjustment – in analyzing industrial location they must be resituated from the center to the outskirts of the model. They cannot determine the course of industry, as in neoclassical theories where exogenous factor endowments and consumer demand fix the coordinates of the space economy. This is equally the case for Weberian industry location theory with its triangulation of transport costs, Ricardian trade theory with its uneven national resource endowments, or the theory of regional equalization in which initial advantage is overcome by the attraction of lower prices at the periphery. On the contrary, prices and incomes are produced by industry over time, along with factor supplies, and concentrations of buyers, in definite places. The result is an ongoing interaction between the production system (and the market signals it gives off to guide the day-to-day behavior of its many participants), and the ability of a dynamic industrial economy to break through existing confines, leap over price and quantity barriers and do things previously unimagined.

Massey (1979, 1984) argues that regional resources are the sedimented results of sequential "rounds of investment," and in this we concur. Industries looking for sites in the present must confront the industrial past in the form of produced resource endowments, often several layers thick. Nonetheless, the metaphor must be an active one in which investment does not merely lay down a fine sediment of effects, but triggers rounds of productive build-up and even sharply disjunctive processes of industrialization at different periods.

3.6 Conclusion

In sum, the industrialization process, with its increasing returns over time, drives location, freeing industry in crucial ways from the prison of the past and the given distribution of factors of production. At the same time, industrialization creates its own markets and inputs to a large degree in a manifestly spatial fashion by building up new industrial centers and peripheries, then abandoning them in pursuit of further opportunities. One cannot imagine industrialization that is not geographical in essence as well as in effect. Industries can develop regions without close adjustment to prior price-quantity conditions; their non-adaptive, production-centered behavior is the primary cause behind the opening up of new locales of production, the formation of dense industrial clusters, inter-regional growth transmitted through dispersal of growth peripheries, and the processes of deindustrialization and reindustrialization that accompany shifting industry centers. Geographical industrialization is not something moving down a predictable path, as in models positing a stable urban hierarchy or the gradual disappearance of regional disparities through spread effects. Events in particular industries and in the short term can realign the development of the space economy in ways that are neither consistent with past trends nor reversible. Industries can extend the centers of territorial development, reshuffle the configuration of a territorially-differentiated economy, and shift the macroeconomic locus of growth by using the microeconomic capabilities available to reasonable actors operating in a disequilibrium economy.

4 Technological Change and Geographical Industrialization

4.0 Introduction

Technology is the magic word in today's ideological lexicon, and with good reason, for the advances across a wide front of "high-technology" endeavors could not be more striking. Rapid technological change offers the promise of salvation to industrialists and presidents alike, from the miracle fiber Kelvar to the miraculous claims of Star Wars. Not surprisingly, technology has come in for intense scrutiny in recent years, more than perhaps at any time since the days of Ure and Babbage at the height of the Industrial Revolution, social science has rediscovered that technology matters.

In this chapter, we shall show how industrial technologies shape the patterns of geographical industrialization we identified as localization, clustering, dispersal, and shifting centers, and thereby play a central role in creating upheavals in the industrial system and in producing persistently uneven geographical development. We shall also seek to turn around the conventional treatment of technology as a producer of geography, and ask how the embodiment of technological practice in territorial production milieux critically influences the social production of technology.

Prevailing models do not do justice to the richness of the relation between technological change and territorial development. Most models may be broadly characterized as "Schumpeterian" owing to their family resemblance to the work of Joseph Schumpeter (1934, 1939, 1942). They rest on three key ideas. First the trinity "invention–innovation–adoption," in which new technologies appear on the horizon, are introduced by leading entrepreneurs, and then diffuse to industrial imitators. Second, the special role of research and development, which gives large corporations the ability to generate their own inventions through captive research laboratories and thereby maintain a persistent technological lead. Third, industrial maturation, or the gradual exhaustion of the technological possibilities inherent in a new invention. The dominant theories of technology and location derive from these precepts (Malecki, 1981, 1983). One such model is innovation diffusion, which claims that innovations arise in industrial centers, especially centers of research and development, and spread from there to backward peripheries. Another model focuses on science-led growth, and makes a case for the uniqueness of "high-technology" industries in regional development. A third model is the product cycle, wherein industries disperse from their historic core areas as they mature.

Schumpeterian thought is characterized by a series of dichotomies: between the realm of science from which invention springs and the workaday world of industry; between leading cores and lagging peripheries; between major and minor innovations; between product innovation and process evolution; between large and small firms; and so forth. Unfortunately, what began as reasonable distinctions, such as invention–innovation, have become frozen into irreconcilable dualisms, blocking the way to further insight. Technology, in our view, unfolds through an ongoing process of interaction between nature and human practice, knowledge and application, routine practice and problem-solving, large and small actors, product and process change, and the like.

4.1 Practice, structure, and technology

First we must establish a clear idea of what technology is and how technological change unfolds. Technological change is produced through practical activity and structured by the physical nature of industries. Technology is the art of transforming nature to human ends. It consists of the knowledge, equipment and practices that make human labor work to an effective conclusion. Without knowledge one cannot act, cannot create the means of acting; but one cannot eat with the idea of a fork. Reducing innovations to ideas sunders technology from its material base of tools, products, and people. Conversely, technology may be substantially embodied in material objects, but it cannot be reduced to things alone; we must be aware of "the part of technology beyond the machine" (Macdonald, 1983, p. 26). The unification of knowledge and technique comes through human practice and that is how technological change unfolds. New technologies are not brought forth complete like Athena born from the brow of Zeus; they are generated through the practical application of ideas, by solving existing problems and discovering new ones, and by the continual updating of knowledge in light of practice (Sayer, 1984). We call this the practical mastery of technology, in and through production. To an important degree, then, technological change is produced: industrial production is at once a process of commodity production and of technological transformation.

At the same time, nature is structured by physical laws which may be mastered but not abolished. The physical contours of products and production processes give shape and direction to industries. The material constituents, dimensions, and functions of objects limit what must be done to produce them, and what can be done with them once they are produced.

Conversely, the capabilities of existing techniques open up certain possibilities to be exploited, press us to explore new territory and lead us to solve new problems.

For Schumpeterians, technology is the prime mover of capitalist expansion, and its roots lie outside the economy, in the realms of science and genius. Inventions derive from progress in science before they are adopted by capitalist entrepreneurs as industrial innovations. Fortuitous groupings of major inventions initiate waves of expansion, which eventually peter out as the innovations are fully exploited and exhausted (Schumpeter, 1934, 1939). Later, Schumpeter (1942) came to think that capitalists had harnessed invention more directly by establishing corporate research and development laboratories as industrial technologies became more "science-based" in the twentieth century; individual inventors could not be expected to muster sufficient resources to compete with such monopolies. In our view, science and invention cannot be granted such primacy in explaining technological advance or new clusters of innovation, since they themselves are embedded in deeper social and institutional structures.

Researchers continue to seek to verify Schumpeter's ideas through studies of the timing of clusters of innovation (Jewkes et al., 1959; Mensch, 1979), the relation of industrial research and development to sectoral growth rates (Mansfield, 1968, 1972; Terleckyj, 1980), the role of science-based industries in growth (Markusen et al., 1986), and the general validity of the long-wave framework (Freeman et al., 1982). Their findings have been inconclusive, however. Clusters of innovations do not appear to lie at the crucial turning points of industrial history or at the onset of long waves of growth; indeed, it is hard to pinpoint or date key innovations with any accuracy. No convincing case has been made that research and development effort, in itself, generates faster or slower growth rates. Nor, despite the number of patents awarded to their labs, do large corporations seem to have a monopoly on innovation and its exploitation. Indeed, several key innovations of the modern era, such as the personal computer, radar and sonar, have come out of the proverbial garage. And while such sectors as computers and missiles employ large proportions of scientific and technological workers, so do some older industries such as basic chemicals and machining.

These difficulties arise from the mistaken belief that technological progress can be traced to a particular source (e.g., the invention of the transistor), or individual or place. While creative insights, historic moments and key choices are not to be gainsaid, such a search does not allow for grasping the long-term processes of technological development that enable these distinctive events to occur. Technological history stems from a systematic growth of human technological powers. The brilliant inventive powers of Edison or deForest, the collective efforts of huge research facilities such as Bell and Westinghouse Laboratories, the research budgets of government programs such as the Manhattan Project must not be confused with the underlying technological possibilities they exploit. Mastery of nature is a social process of production and gaining knowledge which intersects with the structures of nature to produce new frameworks of technological practice.

Let us now amplify these claims as we counter two common myths of technological change and their geographic analogs: first, the idea that social practice in technology development can be equated to the narrow activity known as research and development; second, the notion that modern industry has become "science-based" in a way that breaks with the past.

4.1.1 Research and development in the social division of labor

Everyone knows of the immense size of corporate research and development labs especially in dynamic high-technology industries, and of billions of dollars in government funding for research (Freeman, 1982). This has frequently been cited as the basis for claiming that innovations spring forth from research and development laboratories, then move into production. In geographical terms this view holds that the critical impulses of regional growth emanate from centers of research and development, usually diffusing from there to peripheral areas of branch plant manufacturing (Malecki, 1981, 1983; Thwaites, 1983). Yet none of the many studies designed to prove such a causal link makes a convincing case (Mansfield, 1972; Kennedy and Thirlwall, 1972). Quantity of research and development work appears to correlate well with rates of technological change by industry, but there is little evidence that greater or lesser research and development effort, by itself, will alter those rates.

To begin with, research and development work cannot undertake needed projects and generate useful innovations without considerable input from other parts of the division of labor. If the product has a flaw and doesn't sell, the marketing department is in the best position to know the specific problem. If the new polymer keeps breaking as it is formed into fibers, the production department will voice its concern. Freeman, even in arguing for the vastly increased role of professional research and development, is careful to insist that:

> This does not imply the acceptance of the linear model of research and development with a simple one-way flow of ideas from basic science through applied research to development and commercial innovation. On the contrary, there has always been and there remains in the modern science-related industries a strong reciprocal interaction between all these activities. (1982, p. 107)

Macdonald similarly observes:

> While there is probably some linearity in the innovation process, it is far from clear . . . just where the impetus starts and which direction it takes.

> In the full innovative process, the origin and direction of information flow are even less certain. Invention, it seems, may spring from innovation, development from marketing information, applied research from production problems. Thus, consideration of an innovative, rather than an innovation, process makes much less tenable the assumption . . . that invention is the key element which automatically initiates a chain reaction. (1983, p. 31)

Second, innovations do not come out of laboratories fully formed. There is often a lengthy process of learning before the technology is mastered and adjusted to real conditions. Good ideas must be transformed into prototypes, then incorporated into operational production systems; this requires the labor and intelligence of skilled craftsmen, production managers, salespeople and the like. As a result, working production processes are rarely captured in formal blueprints. As one plant engineer complained: "If we waited until the designs were completed, we would never start building" (Piore, 1968, p. 605).

One clear sign of the need for research and development to relate to and penetrate other branches of large firms is the common practice of splitting such work into units, such as marketing, manufacturing, or materials research and development, and basic science, each with a distinct orientation to the rest of the firm (Gold, 1977). It may be divided up by corporate divisions, as well (Glasmeier, 1985). Indeed, substantial numbers of workers involved in technology development can be found spread throughout most firms – over 80 percent of machine shops have someone in this capacity (Rees et al., 1985). Whether they are called research and development workers is really beside the point: officially designated experts are not the only ones with the expertise needed to get the job done right.[1] The programmer setting up a numerically-controlled machine tool, for example, urges the worker who operates it to discover subtle flaws in the programs (Shaiken, 1984). As Freeman comments:

> The extent of specialization should not be exaggerated. Important inventions are still made by production engineers or private inventors, and with every new process many improvements are made by those who actually operate them. In some firms there are "technical" or "engineering" departments or "O.R." [operations research] sections, whose function is often intermediate between research and development and production and who may often contribute far more to the technological improvement of an existing process than the formal R and D department. (1982, p. 107)

In short, if practical mastery of technologies is to be achieved, technological capability needs to be widely situated throughout the production system.

[1] Managerial insistence on excessive division of labor and compartmentalization of knowledge and work can interfere with the sensible interaction needed for technological competence and innovation. This appears to be one of the contradictions of Taylorism and Fordism (Macdonald, 1983).

In fact, the interaction or inspiration essential to important innovation can occur outside the large firm and research and development labs altogether. Even discounting much of the mythology of "Silicon Valley Fever" (Rogers and Larsen, 1984), it is true that (as with personal computers) tinkering in odd places triumphed over organized corporate or university research and development (Freiberger and Swain, 1984). This does not mean that hands-on experience is more authentic than theoretical knowledge, nor that garages are the true site of invention. Rather, both these types of activity are enabled and supported, as was the invention of the personal computer, by a broader technological milieu in which it is usually impossible to point to a single privileged locus of innovation. The story of Silicon Valley, for example, is one of personal genius embedded in a fundamentally social process of the accumulation of practical knowledge (Saxenian, 1985).

The rise of formal research and development is part of a general process of developing the forces of production. The implementation of more sophisticated technologies has, over time, required an increasing division of labor, and has brought white-coated armies of technicians more directly into the sphere of production to manipulate new materials, design more advanced machinery, and operate automated processes. Some of these people are more theoretically-minded, others are based on the shop floor – but all partake of a broader process of social production.

4.1.2 Practical mastery of nature and technological advance

Technological change takes place through linked sets of inventions and innovations. These sets represent the culmination of fundamental advances in social knowledge and practical mastery. The early Industrial Revolution rested on major steps in understanding of soils, thermodynamics, hydrologic flow, metallurgy, and the principles of making simple mechanisms to imitate hand motions. In the second half of the nineteenth century, there were dramatic leaps in metallurgy, metal cutting and forming, and in extracting organic materials from crude coal. The science-based industrialization of the twentieth century has involved, in the broadest way, a move beyond the mechanical revolution of the nineteenth century, in which the key was manipulation of things (chiefly solids such as wood, metal, and grain) at a level immediately available to human senses and manual action, to a materials revolution involving the ability to command both the atomic and subatomic inorganic structures of nature, and biochemical and genetic processes.

In every case it is not a single invention that causes a technological shift, but a breakthrough into a new technological framework brought about by a whole series of greater and lesser improvements. Some of these frameworks operate locally in particular industries, others provide the base for a wider path of industrial advance. We distinguish the current period as the "age of electronics" rather than "the age of frozen yogurt" because the breakthrough into electronics has opened a wide arena of production and consumption; yogurt technology has a limited place in nature compared to mastering flows of electrons. Similarly the ability to synthesize organic chemicals from oil has vastly greater impact than frozen orange juice because it involves mastery of a whole realm of large-molecular structures, not just a simple freezing process, and permits the creation of a range of associated products and productivity advances in many sectors.

In the process of achieving technological mastery of nature, theoretical knowledge has not always preceded progress in industrial production. Instead, three types of advance – understanding of principles, creative acts, and productive capabilities – have proceeded more or less in tandem within broad technological frameworks. In fact, industrial innovation has often run ahead of the scientific understanding of the underlying principles involved (Rosenberg, 1982, Sahal, 1981). Such practical breakthroughs have challenged scientific minds to come up with better theories and new research programs; the steam engine and thermodynamics is the most famous instance. Precision manufacturing has also been a prerequisite for the creation of most instruments of scientific investigations, from Galileo's telescope to the cyclotron. On the other hand, science has been imbricated in industrialization from the beginning (Musson and Robinson, 1969). For example, the Germans dominated the dyestuffs industry in the nineteenth century thanks to their early mastery of coal-tar dyes through organized laboratory research. By the turn of the twentieth century, scientific research had produced the first synthetic fibers and plastics, revolutionized chemical processing, laid the basis for wireless communication and thermal cracking of petroleum, and even affected metallurgy (Freeman, 1982). Yet even today, despite the advances in understanding of the abstract principles of aerodynamics, materials science, and so forth, the design of a workable jet airplane still demands tedious trial and error and long hours of testing with mockups and prototypes – a process that will never end as long as any failures appear in operating aircraft.[2]

[2] This experience can only be grasped in terms of different types of knowledge and the practical basis of all knowledge. On the one hand, scientific explanation differs from the kind of working technical knowledge generally known as "engineering." Theoretical science seeks to understand the fundamental natural mechanisms beneath commonplace circumstances and events; this means abstracting from intervening causes of particular situations and outcomes. But to make industrial products and processes work, abstract principles must be put into action in real circumstances: it is one thing to know how gravity affects the fall of the apple, another to design an apple picker that works in the wind and rain. Engineers are usually familiar with abstract principles, but they must know more. In practical engineering, "[t]he gist of science lies in indicating what is not possible", rather than determining what exactly should be done (Sahal, 1981, p. 162). Applying scientific principles involves a difficult interplay with the practical capacities of materials, machines and workers; moreover, work may be accomplished with a "practically adequate" working knowledge of things even though scientific insights are still poor. As a result, the performance characteristics of techniques in place cannot be predicted well by theoretical science (Rosenberg, 1982). It is not a matter of "knowledge" versus "practice" but of types of knowledge with different levels of specificity.

To be sure, all modern industry is science-based, if this connotes the systematic application of mechanical, chemical or electronic principles to production, rather than relying on workers' traditional knowledge about their craft (Marx, 1867; Rosenberg, 1976). But the crucial fact of modern industry is not the rise of science to a position over production; rather it is the growth of technology expressed as both knowledge and practical capabilities, over a century of full-scale industrialization and scientific activity. Industrial methods have become more technically sophisticated and exacting but not any less practical for all that, even with the shift from a sawdust-covered workbench to carefully controlled "clean rooms." There is, therefore, no reason to treat high-technology sectors as fundamentally different from other industries in terms of the way technical change affects their locational dynamics.

4.2 Localization, clustering, and the mastery of technology

Technological distinctions between industrial products and production processes, even within industrial sectors, help account for the specialization of regional economies. Traditional Weberian location theory comprehends this spatial differentiation in static terms of varying input–output coefficients: wool inputs to carpets, iron inputs to steel, silk cloth to fashion garments, and so forth. But this does not suffice where inputs are widely available, markets extensive, or buyers and suppliers have grown up alongside an industry precisely to exploit their mutual dependencies. We have argued that industries are capable of developing places by creating their own inputs and markets as they grow, but we did not adequately explain why they should need to do so. The answer lies in the mastery and development of the particular technological bases of each industry, which makes industries distinct in inputs, skills, and markets (Dosi, 1984). One underrecognized source of the differentiation of regional economies is the localization of specialized technological knowledge and its embodiment in the organizational structures of particular firms. This is an intangible but central source of localized and uneven industrial growth.

4.2.1 Natural trajectories and divergent sectoral development

Industries follow divergent natural trajectories, or growth paths, that are structured by the natural bases of their materials, products, and methods (Sahal, 1981; Nelson and Winter, 1982; Dosi, 1984). In terms of production, consumption and evolution, ball bearings are fundamentally at variance with carpets; each industry uses different materials, employs different machines and labor processes, and meets completely different needs. Moreover, the path of technological progress is at right angles in the two industries: advances in metallurgy, machining, lubricants, moulding, etc. hold little direct promise for making a better carpet. Indeed, the practical application of technologies virtually forces certain problems upon users and suggests various solutions, which lie in the nature of the technology itself. Rosenberg calls this chain of events "compulsive sequences" of innovations (1976), and it propels industries down specific pathways depending on their own technological foundation of products and processes. These sequences are based on the relatively fixed technological relations (linkages, complementarities) between parts of products, machines, unified production processes within a factory, or even different sectors of the social division of labor. An improvement in one area can reveal inadequacies in another, as when miniaturization so reduces computers in size that the conventional display screen becomes a barrier to portability, or when expanding memory capacity allows the use of more elaborate software. Such an imbalance may also appear as bottlenecks in the flow of materials from one process to another. There is no reason to expect that such imbalances can ever be eliminated.

Both the rate and level of technical advance are constrained by the technological basis of the industry – there is a world of promise for advances within some industries, much less within others. This is why rates of technical change differ so widely and persistently across sectors, why some processes are much more mechanized than others, and why some products find many uses (large markets) and others few (Kendrick, 1961, 1973). The growth rates of industries are profoundly affected by their technological substrate, as argued in chapter 2; spending more on research and development, for example, will not yield added growth for the industry as a whole if its technology holds few possibilities of further development.[3]

Technological change also spreads through the industrial system in an uneven fashion. Advances in machine tool technology in the late nineteenth century spread far and wide among machinery makers (Rosenberg, 1976); petrochemicals came to supply a wide range of industries from fertilizers to plastic moulding after World War II, and electronics technology is now widely felt (DeBresson and Townsend, 1978). Such technologically related industries, or "technological clusters" (DeBresson, 1987), rest on what may be called base technologies. They are rooted in fundamental technological advances that open up whole new arenas of production and new possibilities in older lines of industry. Wood and cast-iron machinery allowed for only limited precision, speed and strength compared with the

[3] The inability of economists to explain industry growth rates by research and development effort is explicable if the two variables together depend on the possibilities inherent in the different industries' base technologies. As Sahal remarks, "the inventive performance of an industry is determined mainly by the nature of its technology" (1981, p. 57; see also Phillips, 1971). Promising industries have both high levels of research and development and high rates of technical change; hence structured research and development is highly concentrated in a handful of technically vigorous industries (Mansfield, 1972; Freeman, 1982).

steel alloy machines made possible by late nineteenth-century advances in metallurgy and metal working. Petrochemical synthetics brought forth a new world of inexpensive products of magical qualities. Technological space is discontinuous (Perroux, 1950).

As we noted earlier the possibilities inherent in a technology can only be discovered and exploited through industrial production and growth, or what students of technological change have dubbed "learning-by-doing."[4] The idea of natural trajectories adds a twist to practical mastery or learning because "the process of learning tends to be technologically specific" (Sahal, 1981, p. 37). Technology evolves through the process of human exploration of the possible. In the process, every technological choice circumscribes the course of further development by eliminating certain possibilities and by building technique and experience down a particular pathway (David, 1975).

4.2.2 Place-bound and geographically divergent technological practice

The practical mastery of technology is not a placeless process. It is geographically localized by virtue of the particular people, materials, machinery, and firms that embody the experience of an industry. Workers are bound to places by homes, family ties and friendships; firms by investments, faithful customers, and trusted employees and supplies; machines by the bolts in the floor. Technologies are spatially grounded by all these local ties, but also by the mutual play of competence with specific kinds of products and labor processes across the social division of labor in every industry. Industrial locales provide a **technological milieu** of great richness, going beyond access to information (Pred, 1974) to include both the accumulated know-how, and active techniques of industrial production. Thus, technological discontinuities – whether from single-sector natural trajectories or from multi-sectoral base technologies – can manifest themselves on the industrial landscape, in such notable clusters as the Midwestern metal-working complex, the Los Angeles aerospace complex, or the New York financial complex. Figure 4.1 shows the linked systems of firms that developed over time to create the machinery and electronics clusters of Massachusetts in the nineteenth and twentieth centuries.

The importance of specialized suppliers, buyers, and workers in binding firms to a particular location goes deeper than market size, economies of

[4] For a review of the literature, see David, 1975. Industries have even been found to enjoy steadily increasing worker productivity without any major change in technique, as a result of worker experience with, and perfection of, existing methods alone (Arrow, 1962). This has been called "disembodied" technological change, because it involves no investment in fixed capital; but the distinction is false because learning does not affect just the worker, it has an impact on the installed technique: altering machines to get them to run correctly, adjusting production lines to operate smoothly, reorganizing the division of labor and allocating work load efficiently, and so on.

scope, or price; behind those lie the command of specific technologies that gives content to such abstract categories as skill, cost functions or product demand.

We mean more here than the conventional, static notion of access to specialized labor, suppliers, or information. What is wanted is not an off-the-shelf product or labor-service, but someone or something to solve an indeterminate problem: the seamstress who can turn a radical design into a fitted dress, the specialized machine tool that can cut a certain pattern in metal, the programmer who can come up with the right subroutine, the metallurgical team that can create a new steel alloy. What is usually called for, moreover, is not one-time access but ongoing involvement with suppliers, consultants and buyers, with considerable back and forth communication and movement of personnel. Technological localization is thus a process of learning by problem-solving as an industry or closely-linked group of industries grow and change. Industrial complexes, by bringing together so much activity, of such variety, not only increase the frequency of contact and sharing of knowledge, talents, problems and demands; they increase the probability of hitting on significant innovations, and of allowing their rapid spread and embellishment through the network of firms and activities. Industrial localization is an outgrowth of organizing production under conditions of technological flux; conversely, place-bound technological practice contributes to the formation of industry and place-specific technological trajectories.

Capitalism, of course, is an unparalleled force for the homogenization of productive activity and social life around the world; this uniformity reflects, among other things, the globe-girdling flight of commodities, the impress of market-rationality on diverse cultures, the installation of wage-labor, and the utilization of similar production methods, such as the assembly line, throughout far-flung corporate dominions. Competition, both inter-firm and international, forces everyone's hand, and makes them play by common rules to survive. There is immense pressure, for example, to build TVs and VCRs like the Japanese, supercomputers like the Americans, sporty cars like the Italians, or household appliances like the Germans. So cars, coffee grinders, and airplanes, and the machines and methods used to make them, come to resemble each other. But they are never strictly identical, except where the original company builds branch plants or the hapless competitor licenses (or copies) the technology from the leader firm. Yet behind the wave that seemingly sweeps all away come others that re-establish uniqueness and difference. One source of difference is that industries are technically heterogeneous; another is that even the same generic industry can evolve along variant trajectories according to local practice based on long experience with specific products and processes. American Midwestern farmers are very good at land-extensive cultivation, for example, because for over a century farmers and machinery makers have worked to maximize yields with a family-based workforce

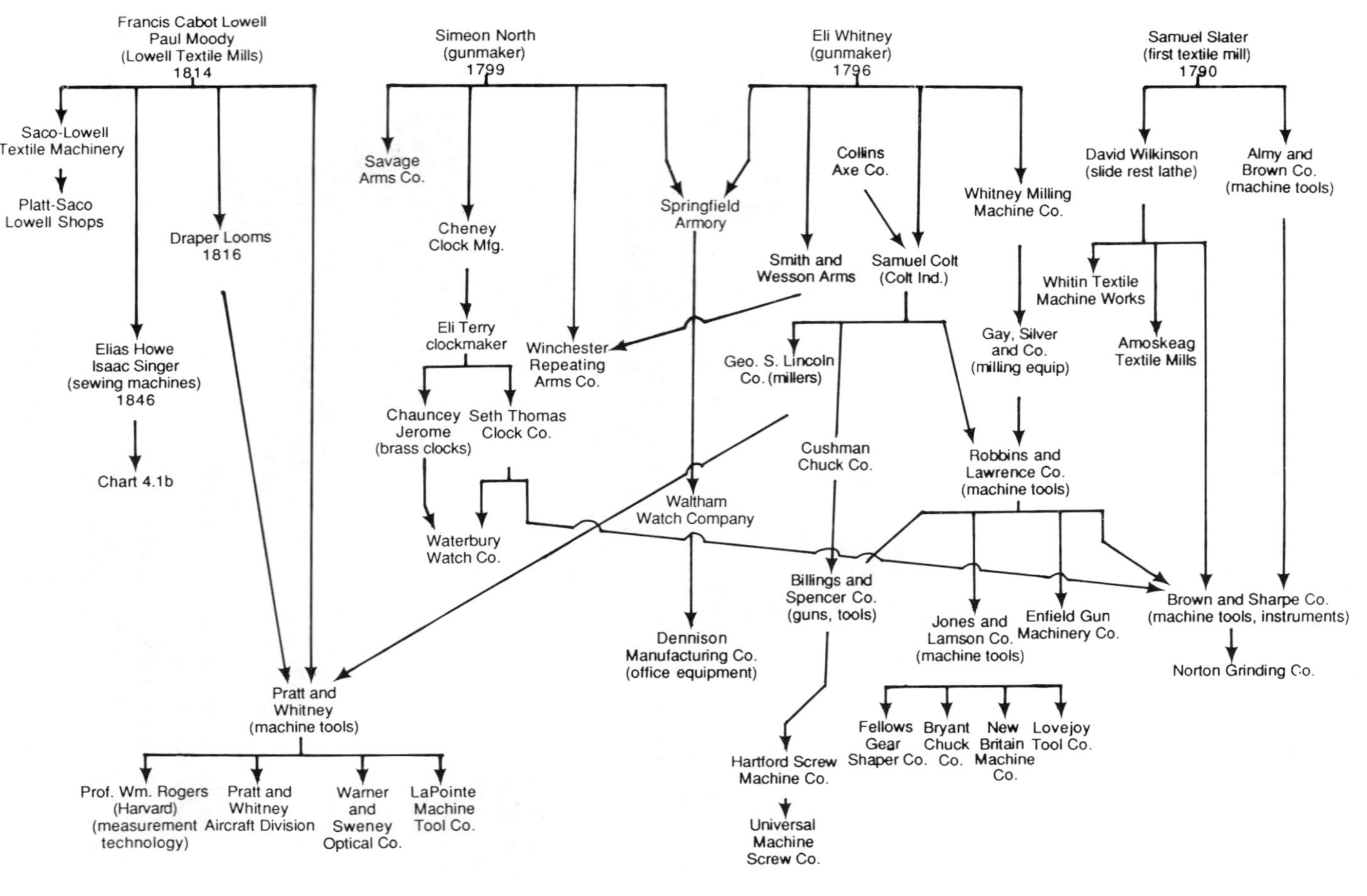
Francis Cabot Lowell
Paul Moody
(Lowell Textile Mills)
1814
Simeon North
(gunmaker)
1799
Eli Whitney
(gunmaker)
1796
Samuel Slater
(first textile mill)
1790
Saco-Lowell
Textile Machinery
Platt-Saco
Lowell Shops
Draper Looms
1816
Elias Howe
Isaac Singer
(sewing machines)
1846
Chart 4.1b
Savage
Arms Co.
Cheney
Clock Mfg.
Eli Terry
clockmaker
Chauncey
Jerome
(brass clocks)
Seth Thomas
Clock Co.
Waterbury
Watch Co.
Winchester
Repeating
Arms Co.
Springfield
Armory
Waltham
Watch Company
Dennison
Manufacturing Co.
(office equipment)
Collins
Axe Co.
Smith and
Wesson Arms
Samuel Colt
(Colt Ind.)
Geo. S. Lincoln
Co. (millers)
Cushman
Chuck Co.
Billings and
Spencer Co.
(guns, tools)
Hartford Screw
Machine Co.
Universal
Machine
Screw Co.
Whitney Milling
Machine Co.
Gay, Silver
and Co.
(milling equip)
Robbins and
Lawrence Co.
(machine tools)
Jones and
Lamson Co.
(machine tools)
Enfield Gun
Machinery Co.
Fellows
Gear
Shaper Co.
Bryant
Chuck
Co.
New
Britain
Machine
Co.
Lovejoy
Tool Co.
David Wilkinson
(slide rest lathe)
Whitin Textile
Machine Works
Amoskeag
Textile Mills
Almy and
Brown Co.
(machine tools)
Brown and Sharpe Co.
(machine tools, instruments)
Norton Grinding Co.
Pratt and
Whitney
(machine tools)
Prof. Wm. Rogers
(Harvard)
(measurement
technology)
Pratt and
Whitney
Aircraft Division
Warner
and
Sweney
Optical Co.
LaPointe
Machine
Tool Co.

Figure 4.1(a)

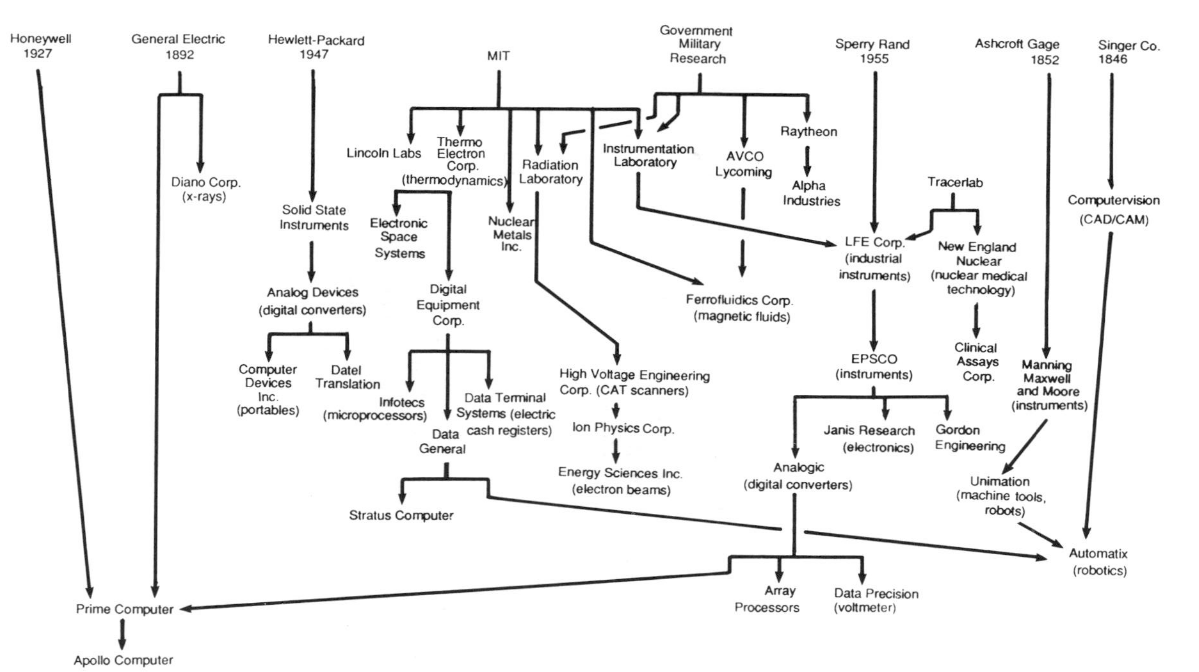

Figure 4.1(a,b) *Technological continuity in New England, 1790–1900 and 1900–1980* Any successful industrial center spawns a succession of technologically interconnected activities, linked by movement of personnel, continuity of firms, spinoffs of new companies, and common university or government ties. The evolution of New England's textile and machining complex of the nineteenth century and electronics complex of the twentieth century provides a graphic illustration of the principle. There is little continuity between the two complexes, although a new grouping of electrical firms, university research, and military funding indicates that a watershed has been crossed. (After Hekman and Strong, 1981)

and to extend holdings. But for yield per acre, US farm technologies can't hold a candle to the Dutch, who have been squeezing the most out of a little space for centuries. The French still make the best cooking utensils, by and large, and have swept the kitchens of the world with their food-processors; but for coffee-making equipment, look to the Italians, with their great coffee culture. Even across a supposedly universal industry such as automobiles, one still finds a variety of practices of design and manufacture between the US, England, Japan, or Italy (Friedman, 1977; Abernathy et al., 1983; Cusumano, 1985).[5]

Nor is the divergence confined to individual commodities. English and American technologies diverged across a broad front in the nineteenth century. American machinery was more standardized, simple and powerful, less well crafted, and more widely adaptable; in general, US technology was more mechanized and more wasteful of natural resources (Habakkuk, 1962; David, 1975; Vance, 1986). The most controversial comparison today is between the United States and Japan. The Japanese produce better quality products at a lower price across a wide range of industries, especially automobiles and consumer electronics, but also steel, ships and some semiconductors; and they do so through a set of innovative methods of production (Schonberger, 1982; Okimoto et al., 1984; Cohen and Zysman, 1987). Yet US manufactures still have areas of strength in both old and new sectors – from aircraft to computer-aided design (CAD) software – which the Japanese have yet to emulate or transcend.

Technological divergence is also prevalent between sub-national regions, but this has received little attention in the study of industrialization. A close reading of the metal-working industries of the United States reveals a definite split between the "American system" of New England, which emphasized machining to close tolerances, experimenting with interchangeable parts and using machines in place of hand labor, and midwestern practices, where machining was less artful and sheet-metal was often worked to arrive at the same results. For example, in New England a bicycle hub would be cut down from a solid steel bar, whereas in the Midwest it would be built up from a piece of sheet metal (Hounshell, 1984). Part of the genius of Henry Ford and his associates was to marry the best of both traditions into a new one, a classic way of achieving a technological breakthrough (Sahal, 1981). California agriculture has long been very different from the dominant family farm system of the eastern and midwestern United States, owing to its larger landholdings, heavier

[5] Because product and process heterogeneity persists, we are presented with the irony of Parisian suburbanites occupying California-style ranch houses built by Los Angeles homebuilder Kaufman and Broad while stylish Los Angelenos are buying French clothes in record quantities. While such internationalization of consumption may look like homogenization from one perspective, from another it looks like the arrival of something new and different.

capitalization and large-scale use of immigrant labor; this has affected farm technology with California being an innovator in such things as irrigation, the caterpillar tractor, and the tomato harvester. Historically, the diffusion of technology among regions has been as important as that between countries, even though it has attracted much less attention (Pollard, 1981).[6]

Localized technological change in an industry can be understood, like all industrial development, as an evolutionary path in which each step moves one away from a past that cannot be recovered and that limits future directions (Nelson and Winter, 1982). Yet spatial differences in industrial practices are conventionally explained as an efficient equilibrium allocation of industry based on resource endowments and consumer tastes, given by nature or by an unexamined history yielding an invariant "culture." The Italians have a strong small-firm craft tradition and a delight in good coffee; the French a more centralized, bureaucratic tradition and a taste for fine wine; the US a lot of grasslands and a love for beef; the Dutch a small country won from the sea and an affection for herring; and so on. These endowments are supposed to account for their noticeable industrial and agricultural differences. But this is *post hoc* rationalization of a long process in which skills, consumer tastes, machines, dikes, drip irrigation systems, and so forth develop through an interactive process of learning-by-doing and learning-by-using (Rosenberg, 1976). Endowments are creations of local technological history as a part of local industrialization as previously argued (section 3.5). Their step-wise and irreversible growth process cannot be captured in terms of continually weighing a field of alternatives in terms of relative prices. Indeed, without a lengthy process of developing a particular practical capacity, one usually cannot choose to employ that technology at all, regardless of factor endowments and relative prices, and once the capacity to produce is lost, all the cheap labor or tax incentives in the world cannot quickly bring it back again (Cohen and Zysman, 1987).[7]

[6] Differing place-specific technological practices can also lead to minor product and process divergence. For instance, localized technological capability has been a vital source of product differentiation among US, Swedish, German, Italian, and Japanese automobiles. There are subnational examples as well, as in the case of competing furniture industries surviving in North Carolina and Los Angeles. As a result, multiple locations can persist despite the dominance of one center – competition does not drive out all contenders, as in the simple model enunciated in chapter 3.

[7] In semiconductors, for example, US independents made a fateful turn when they moved assembly of standardized chips (primarily RAMs and EPROMs) to Southeast Asia instead of mechanizing. As a result these firms lost the chance to learn the capabilities and problems of such machinery, experience which had an increasing payoff as the scale of chip markets grew. Japanese companies, meanwhile, worked to perfect mass-production techniques though they lagged in chip design, and succeeded in conquering their US competitors in the 1980s. In response, Silicon Valley firms emphasized their advantage in chip design of advanced integrated circuits by moving into specialty markets and subcontracting fabrication to Japanese firms. The United States thus appears to be falling farther behind in production capabilities while maintaining its lead in design technologies.

4.3 Place-specific innovation and the rise of new industry centers

Technological practice does not reinforce the process of industrial localization indefinitely. Technological ruptures can provoke industry relocation and recentering. Place-specific technological advance may also generate superior practices that alter the competitive standing of national or global centers of industry.

4.3.1 Technological breakthroughs and locational shifts

Given the structured nature of technological change, it is possible to speak of technological breakthroughs, or qualitative shifts to new technological frameworks that put industries on a new footing. Breakthroughs may consist of a new product design such as the DC-3 aircraft, a new production method such as the Bessemer steel process, or the use of a new material base such as ceramic superconductors. Paradoxically, breakthroughs do not ordinarily occur in one epochal invention or single act of genius, but become apparent as a line of development reveals its technological potential, often as the unanticipated result of solving a smaller problem – "overshooting of the mark is characteristic of exploratory activities . . . [T]he size of the discovery need bear no systematic relationship to the size of the initial stimulus" (Rosenberg, 1976, p. 115). In other words, the evolutionary paths of technological development may reach dramatic, qualitative breaks that lead to temporal kinks in the technological histories of industries. The carriage became the automobile with the introduction of the internal combustion engine, and while carriages and cars share some important characteristics of size, use, and components, the automobile rapidly sped away from its predecessor.

The possibility of technological change altering the locational logic of industries is understood by writers in the Weberian tradition, but the historical element has been relegated to an afterthought (e.g., Hoover, 1948; Isard, 1956). Periodic ruptures of production generated by breakthroughs in technology make for disjunctures in spatial development. There are two principal geographical manifestations of kinked technological paths. Technological change may so alter the nature of the product that a completely new industry is defined, and production technology must be largely generated *de novo* – as when the harvester industry broke away from the rest of farm implements in the 1850s (Pudup, 1987). As inputs were not available, the new producers enjoyed a wide locational choice. The centers of older industries may also shift because technological changes redefine the opportunities for large-scale or agglomerated production and so induce creation of new centers, or because they transform product form or production processes significantly but still within the industry.[8] Meatpacking provides a good example of the importance of such technological shifts.

The meatpacking industry, properly so called, was constituted with the transfer of hog slaughtering from farm to factory in the early nineteenth century: the introduction of the disassembly line and large-scale hogpacking (in brine), orchestrated by merchant capitalists, led to a concentration of the infant industry around Cincinnati in the 1830s (see figure 3.9). Westward farm expansion and the shift from canals and steamboats to railroads helped move the axis of production to the Mississippi Valley in the 1850s. Chicago's pre-eminence was secured, just after the Civil War, by a major organizational innovation: the creation of the stockyards by the railroads. These made Chicago the focal point for shipping beef on the hoof to eastern markets. A further epochal shift that made Chicago "hog butcher to the world" was Gustavus Swift's refrigerated freight cars of the 1870s. Animals could now be slaughtered in the Midwest instead of being brought to eastern markets, and midwestern factory slaughter-houses were able easily to outcompete eastern workshop-scale butchers. With factory-production, packing soon consolidated around the Big Four packers and the industry settled into a clustered oligopoly by the turn of the century. After World War II, however, a new period of restructuring began. The modern feedlot, with its technologies of breeding, mass feeding, inoculation, and hormone regulation agglomerated the production of cattle themselves. Slaughtering moved farther west to be near the high plains feedlots. Central to this shift was "boxed" beef (pre-cut and pre-wrapped in vacuum plastic) in place of "dressed" beef (carcasses or large slabs) introduced by Iowa Beef Packers (IBP); significant improvements in factory mechanization, especially wrapping and boxing methods, made this change possible. Boxed beef could go directly into grocery stores, undermining organized retail butchers and enhancing factory production. IBP now commands almost 40 percent of the market, and two other new entrants to meatpacking, ConAgra and Excel (owned by Cargill), another 40 percent; little remains of the companies for which Chicago was once famous.[9]

[8] For a shift in an industry's center to occur, however, the underlying requirement is for sufficient technological change to overcome or circumvent the agglomeration economies in existing centers (see chapter 5).

[9] Labor relations also played a role in the shift out of Chicago and places such as East St Louis, which became unionized in packing during the 1930s. Company antipathy to militant workforce – especially to black and Communist workers – was strong from the late 1940s on, and hastened relocation even before Iowa Beef Packers appeared on the scene. IBP is fiercely anti-union, and situated its plants in small towns distant from any labor tradition. Thanks to Brian Page and Mary Beth Pudup for explaining the meatpacking story to us.

4.3.2 Divergent technologies and shifting centers

Divergent technologies are not merely a source of geographical heterogeneity: they are frequently the crux of competitive advantages allowing firms from a particular national (or regional) base to rise up and conquer the world of commodity production. We must consider the geographical basis of technology as part of the process of spatial restructuring, including both the challenge to old centers of industry presented by new ones, and the way territories contribute to their own industrialization and deindustrialization.

At one pole is the exhaustion of industrial growth in established centers: aged heartlands often fade into obscurity as have Massachusetts' shoe producing towns, or are completely extinguished as was Coalbrookedale and its iron forges on the Severn. These cases of regional stagnation cannot be ascribed simply to declining product markets or plant dispersal, for at the other pole are new industry centers which outcompete old ones by virtue of better products or more efficient processes, taking away markets and forcing cost-cutting strategies. There are potentially many new reasons for shifts in industry centers, but here we concentrate on only one: the role of divergent place-specific technological practice.

Entrenched practices and established trajectories often appear to have a deadening effect on industrial change in established production centers. The principal reason for this is not that a local industry and its firms have become technologically incompetent over the years, rather that they are, if anything, too good at what they do. They need an injection of a new technological and organizational approach brought by an unconventional marriage of techniques, and a general ability to experiment with possibilities that do not fit with old habits.[10] Breakthroughs into new technological frameworks and growth trajectories usually come about in new places. Ironically, a major innovation may be born in an established industrial center, but undergo its full development elsewhere – creating a new technological milieu and industrial environment in the process. In many cases, ideas – and the workers who fought for them within existing companies and lost – depart established centers to work unimpeded on their own. William Shockley and others headed west from Bell Labs in New Jersey to found Shockley Instruments, Texas Instruments and Motorola (Dosi, 1984). Ken Olson left IBM in Poughkeepsie, NY, to form Digital Equipment (DEC) near Boston. The movie moguls of Hollywood departed New York, where they chafed under the dominance of Adolph Zukor, to establish a new center for the movie industry that eventually outshone the old one.

[10] We are not, of course, discounting labor relations, politics, or resource exhaustion as sources of decline (see chapters 6 and 8).

But how can technically backward regions ever become competent and innovative enough to compete with existing industrial centers, and even surpass aging regions in technological dynamism? This is especially puzzling in cases where there has been no obvious transplant of companies, divisions or major personnel. The United States and Germany surpassed Great Britain on the strength of such autonomous innovations as the "American system" of standardized parts and the German system of technical universities and scientific chemistry (Schumpeter, 1939; Hounshell, 1984). The technological achievements of German and American industrialization, and of Britain in the eighteenth century, rested solidly on each country's experience and creativity. The successful industrialization of every periphery has its own history, and a lively one at that. Long before the Midwest became the automobile center of the world it had an industrial base: farm equipment, meatpacking, grain milling, carriages, bicycles, ball bearings, general machining and metal working, and Pullman cars. Long before anyone had heard of high-technology industry, California had been manufacturing pumps and mining equipment, making explosives, pumping and refining oil, drying and canning fruit, and assembling cars and trucks.

Hierarchical diffusion models founder on the question of how to transform the periphery from a passive recipient of innovations and sterile branch plants to a vital center of technological change (e.g., Lampard, 1955; Norton and Rees, 1979). Such models have no grasp on the vital, if struggling, process of industrialization going on at the periphery; instead, technological change flows from the center until a "take off" occurs in the periphery. In our view, the actual practice of manufacturing at the periphery is a fundamental precondition for future advance. Industrialization is a continuous learning process, in which the accumulation of capital goes on at the same time as the accumulation of labor skills, of collective knowledge, of requisite machinery – in short, practical mastery of industrial technology.

Backward places do often begin by making cheap imitations or producing strictly for local (and protected) markets. Taiwan and Korea have followed Japan in beginning with the oldest and simplest industrial technologies, such as textile weaving, then moving into heavy industries such as steel and shipbuilding, and finally entering into competition in newer sectors such as computers and semiconductors (Cumings, 1984). Strict mimicry, taking on the most lowly and banal tasks in the division of labor, and relying on low cost production in older lines of industry is not enough, however. Originality is crucial, for therein lie the new products and processes that give a region competitive advantage. The successful industrializers manage frequently to exceed the mastery of their competitors and make significant breakthroughs, as the Japanese did in steel (Gold, 1977) or Los Angeles did in women's sportswear (Pitman, 1987). Diffusion thus involves more than the passive receipt of information or commodities to be used according to instructions. It is a process of adaptation involving modification to product designs; adjustment and breaking-in of machines and processing methods; introduction of new divisions of labor involving learning, task alteration and considerable haggling among workers and management over just who does what, how, where and when (Morgan and Sayer, 1988). In other words, the use of outside technology can be structured so as to involve problem-solving and thus open up the possibility of new solutions to old dilemmas, as the Japanese did with American automobile technologies in the 1950s (Cusumano, 1985). The Korean strategy with respect to Japanese firms is also instructive. They try to borrow selectively, securing proprietary foreign technology for parts and processing methods they absolutely cannot do themselves, while remaining independent where they can. In the former areas, they work under license until they can master the technique and come up with a new solution which is theirs to use and to sell, in turn (Foster-Carter, 1985; Jacobsson, 1985).[11]

The result, as Gerschenkron (1962) observes, is that backward countries do not follow the same line of development as those that came before them, but trace quite different paths reflecting the collision of their own histories with the forces of capitalist industrialization (Pollard, 1981). Gerschenkron's formulation captures an important dimension of the process of capitalist expansion, but it is trapped in an imagery of time that obscures the equally potent differentials of space. Not only do countries enter the maelstrom of industrialization at different times, they enter at different places, with their own particular societies, industries, and possibilities.

4.4 Technology and industrial dispersal

The prevailing view in both neoclassical and Schumpeterian theories of location is that the main effect of technological change is to stimulate industrial dispersal. Such a view is based on misconceptions of how technological change proceeds over the course of industrial development.

In the terms of Weberian location theory, the principle form of technological change involves improvements in transportation and

[11] A United Nations Economic Commission on Latin America (ECLA) research team found that indigenous technological capabilities have developed most extensively in industrial fields that produce in small batches, rather than in those making standardized products in large series. In the latter, foreign multinationals have a great edge, and much of the crucial technological work of product and process design is done in the home country, before the line is set in motion. In the former, by contrast, more of the engineering is done through *ad hoc* and on-site problem-solving by skilled workers, providing much greater scope for learning and arriving at original solutions (Katz, 1983). Of course, places so undeveloped that they lack any substantial industry, local or foreign-based, have few possibilities.

communication methods. These progressively lessen the friction of distance, thereby diminishing the force of spatial proximity and allowing industries to disperse their facilities. At the same time, the higher land and labor costs of built-up areas can be avoided at the undeveloped periphery. It follows that there is a natural spread effect in capitalist industrialization with technological progress, and this has been offered to explain the decentralization of industry from inner cities to suburbs, from the Northeast to the Southwest and from developed countries to the Third World (Chinitz, 1960; Borts and Stein, 1964; Kain, 1968). At its most extreme, the belief is that industry is no longer bound to particular places at all, it has become "footloose," free to follow the whim or residential preferences of owners and managers (Berry, 1972).

Such models presume technological stasis in every sector other than transport and communications. This is manifestly not the case. On the contrary, technological change actually often reduces footlooseness. As we have just argued, rapid technological change can make it important to stay in the midst of an industry's technological milieu. Even improvements in transport and communications may have an effect contrary to expectations: the telegraph and railroad actually increased the relative advantages of large factories over small producers serving local markets and of big city locations over small town and rural ones (Pred, 1974, 1980). Similarly, the Wall Street congregation is not about to dissolve due to worldwide telecommunications improvements; rather those same technologies are causing rapid innovation in financial operations and conferring on New York even further advantages as securities capital of the world. This is because the very same technologies that permit ease of interconnection also permit restructuring of markets, and lead to the creation of new specialized functions within the division of labor. In the banking and securities industries, this has led to the proliferation of small and medium-sized firms serving global markets through interaction with other specialized firms. Agglomerative tendencies of the industry have thus been reinforced rather than diminished.

Another important model of the relation of technical change to decentralization is the product cycle, which is of Schumpeterian rather than neoclassical inspiration. Relative spatial factor prices still determine locational behavior, but only as the parameters of production change over time. The product cycle literature is somewhat confusing as it has two emphases, one on product maturation and the conquest of extensive markets, the other on production maturation and the growing standardization of inputs (Hirsch, 1967). The product version argues that new items appear on the market as luxury, novelty or specialty goods, go through a period of market development, eventually expand to become mass consumption goods, and finally decline with the advent of substitute products, market saturation and foreign competition (Dean, 1950; Levitt, 1965). In the seminal statement of this version, Vernon (1966) argued that US companies tend to introduce new products in regional markets, then conquer the American market and finally move overseas; as the product gains a foothold abroad, demand is met by export but as that market grows with product maturity, firms build overseas production facilities (Wells, 1972).

The other version of product cycle theory focuses on maturation of production process technologies. A youthful industry makes small batches of its new product using skilled labor, which is best found in major industrial centers; as it matures, the industry standardizes its product, routinizes its activities, mechanizes production, deskills the labor process, purchases in large stable quantities, and builds large self-contained factories; it thereby depends less and less on economies of agglomerations and can deagglomerate in search of cheaper labor and cheaper land (Scott, 1983). Kuznets (1930) and Burns (1934) originally explored this idea in a non-geographic way. The earliest geographical application of the "industrial retardation" idea appeared soon thereafter (National Resources Planning Board, 1943; Hoover, 1948); but the popularization of a full-blown production cycle model is due chiefly to the work of Vernon (1960).[12]

Product maturation and market spread The market expansion variant of the product cycle argues that a product goes through an inevitable process of standardization with time, and this in turn permits its manufacturers to extend their market territory. Yet, what Vernon interpreted as market extension via the slowing of technological change has been reinterpreted, by those who theorize about the multinational firm, as market extension due to continuing mastery of complex and changing technologies. In this line of thinking, the big firm possesses a specific competitive advantage in its mastery of a set of production technologies so complex and interlinked that changes in one part of the process (e.g., the production of a component) must mesh with changes elsewhere in the production system. The necessary control and integration of the technology is either internalized within the large corporation or carried out by a local enterprise under license, usually with some market protection from the host government. (Buckley and Casson, 1976; Dunning, 1979, 1981; Hennart, 1982; Caves, 1982). If the product maturation version of the product cycle were correct about industrial diffusion, it might be supposed that, other things being equal, local competitors would arise to compete with outside firms in technologically-standardized domains. But that is by no means typical: companies usually capture new markets precisely because they possess product and production technologies that are difficult to imitate. Mastery of technologies is not a matter of patent monopolies

[12] Thomas (1975), Krumme and Hayter (1975), and Erickson (1975) deserve credit for its recent popularity in geography.

alone, but extends to the firm's workers, machinery, and documents. Worker migration between firms or nations, with blueprints for machines in their heads or hands, has been an integral part of industrial "dispersal" over the years, from Samuel Slater's emigration to Rhode Island to Robert Amdahl's defection from IBM in Silicon Valley (Pollard, 1981).

Production process maturation In this variant of the product cycle, production methods become more standardized over time (see figure 3.5). The rapid growth of youthful industries is ascribed, in Schumpeterian fashion, to "a revolutionary discovery or invention" and decline to "the scope for invention [being] gradually exhausted as all important operations are mechanized and perfected" (Kuznets, 1930, p. 198). The chief shortcoming of this conception of technological development has been discussed in the preceding section: industries have considerable powers of restructuring and renewal. They manifest "kinked" development paths as technological practices move from one technological framework to another. That is, technological exhaustion, or maturation within a framework, must be distinguished from technological breakthroughs from one framework to another. Short-term change takes place within a structure, and is incremental; while long-term change involves movement between structures, and is transformational. The product cycle notion that an industry is "born" and then "lives out" its days relies on an erroneous analogy to the individual organism; the evolution of species may be a more appropriate metaphor (Gould, 1977).

The production maturation model also contains an overly simplified perspective on mechanization; every industry begins with manual labor and gradually elevates itself to a level of automated mass production.[13] But automation does not proceed in quite this regular way (Walker, 1988b).[14] First, the possibilities for automation differ among industries owing to their fundamentally different products and material foundations. The mechanization of cement-mixing is worlds apart from the mechanization of electricity generation. Industries may be blocked from becoming highly automated by the size of the product (airplanes), or its shape (garments), by the material used (wood), or the uses of the product (scientific instruments). Only a select group of industries have been able to exploit assembly-line or liquid flow techniques (Littler, 1982, 1985). Second, mechanical progress in any industry does not involve a linear ascent from manual to automated processes, as portrayed by Bright (1958). It advances unevenly along several dimensions of the labor process and among the various parts of large production systems (Bell, 1972;

[13] The relation of mechanization to deskilling is considered in chapter 6.
[14] Much of the dispersal of plants to the Third World in garments and textiles has been due to the inability of these industries to mechanize in the face of strong competition (Jenkins, 1984).

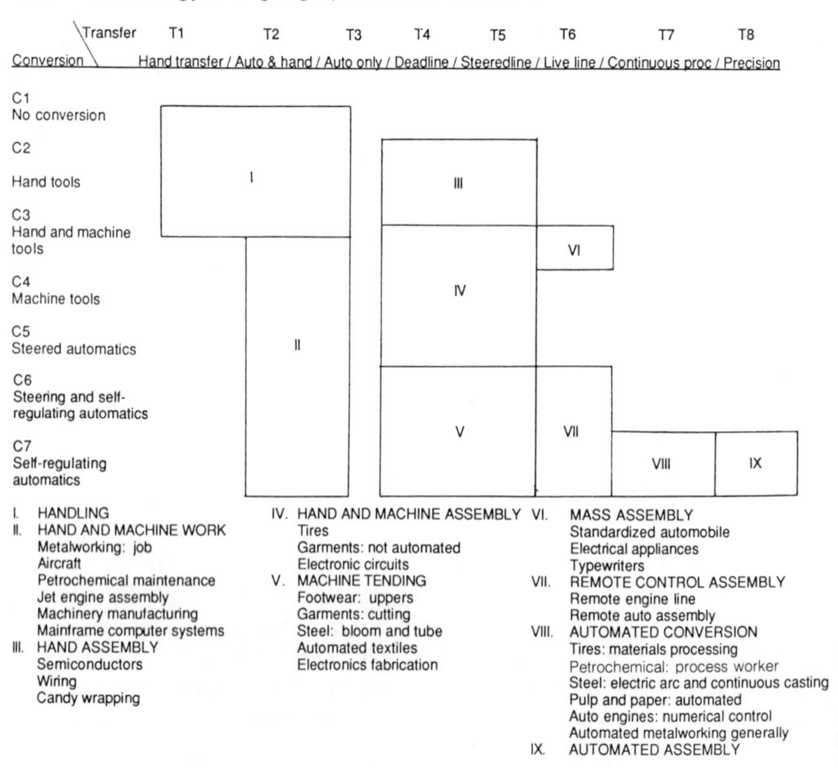

Figure 4.2 *Types of workplace technology* Industrial processes are here mapped according to level of mechanization along two dimensions, conversion and transfer, to show the non-linear development of process automation.

Kaplinsky, 1984). Automation can actually be reversed in one dimension or realm in order to facilitate advances along another. For example, a highly mechanized cement plant may generate a great deal of extra hand labor in bagging, packing, and shipping (Nichols and Beynon, 1977). New innovations in production methods may enter from a side route and overtake the most advanced methods, as with the challenge to Ford's assembly line presented by Japanese car makers' just-in-time deliveries, fewer levels of managerial oversight, "kanban" system of materials tracking, and greater worker involvement in decisions (Sayer, 1986b). The ultimate lesson is that automation is open-ended. Tomorrow's advances in neural network computers, robotics or computer-aided design may make present achievements look very modest indeed, and may reorganize production in unexpected ways.

The unevenness of technological progress is depicted in figure 4.2. It classifies technologies according to the methods they use to convert raw materials on the vertical axis and the methods used to link work stations to each other on the horizontal axis. Technologies tend to cluster at particular points in this theoretical matrix of possibilities; real technologies do not exist for every box, and box IX is at present a poorly-realized possibility in even the most automated production systems. Technological development tends to move from possibility-set to possibility-set in an uneven and somewhat unpredictable way. Moreover, many of the most important contemporary high-technology industries are found in cells II and III of the matrix, while many of the rapidly-growing manufacturing industries – such as clothing and specialized craft industries – are found in cells IV and V. These technologies are by no means disappearing from the industrial landscape of advanced capitalism.

4.5 Conclusion

This inquiry into technological change and geographical industrialization suggests a set of general conclusions. On the one hand, industrialization is technologically structured, even at a very large scale, in such a way as to generate common patterns of localization, clustering, dispersal and shifts across a broad range of industries. On the other hand, geographical differences and place-based practices shape some of the very technological changes that profoundly affect location patterns. By joining the two sides of geographical industrialization as they relate to technological change, we can see more clearly the origins of the major macro-spatial puzzles enunciated in chapter 1: the push into new territory, the differential geography of regions and nations, and the disruption of established patterns of urban-industrial hierarchy.

Geographical expansion is possible in part, because as the technical horizons of industrialization are pushed back, opportunities arise for new firms in new places to achieve practical mastery over previously unconquered domains of production. Technical change does not trickle down the urban hierarchy so much as bubble up from the productive base, giving the existing pattern of spatial development a bed of quicksand on which to rest. Conversely, the commitments of previous centers of industry to their special technologies render them often less rather than more fit to pursue diverging lines of activity. Thus, leading industrial territories do not inevitably stay in the lead, despite the many advantages leadership confers, nor are lagging territories condemned to eternal backwardness, despite their many handicaps. In both instances, the specificities of industry trajectories and even of wide-ranging base technologies are deeply bound to the places in which they develop; their collective technological milieu is not something easily isolated, patented, or transported for diffusion to other locations except by those who participated in the core's development. The place-bounded aspect of technology does not foreclose the possibility of competitive challenge within an industrial sector, however. Divergent technical practices are possible even within the boundaries established by common products and materials, allowing for some range of product diversification, production competency, and localization. More than this, industry development paths are not only subject to technical breakthroughs that can dislodge the primacy of an extant production center; the geographic diversity of technological practices provides a potent spring from which may flow breakthroughs into new product or process configurations. Indeed it is the collision, as well as the division, of technological trajectories that usually stimulates the most fruitful restructurings.

While technology alone cannot explain industry paths and the fortunes of places, the geographical behavior of capitalist industry does appear to be bound to a significant degree, with the rise and fall of divergent technical frameworks and base technologies. We can embrace strong technological determination without falling prey to technological determinism, however. Many questions about geographical industrialization cannot be answered without reference to other fundamental features of capitalist production, such as labor relations and industrial organization. Indeed, the effects of technological change are usually filtered through altered work practices, union negotiations, decisions to subcontract or produce in-house, and the like, making it difficult to link technology and geography directly. We take up industrial organization and labor relations in the chapters to follow.

5 The Territorial Organization of Production

5.0 A unified approach to organization and location

In most geography and economics the industrial system is carved into familiar pieces. Everyone knows what is meant by "the car industry," "global corporations," and "capital-intensive factories" – or do they? Ex-Private Walker, US Army, recalls his surprise on finding his first M-16 stamped "Made by General Motors." Does this mean GM is part of the arms industry instead of the automobile industry? Neoclassical economics and Weberian geography beg the question of organization by treating the market as the mediator of all economic transactions, the plant as a production function, the firm as a single plant, the industry as made up of representative firms producing a single product, and the region as a blank slate on which firms individually pick out the best spots to locate. Faced with the obvious discordances between these categories and reality, students of industrial geography have tried to reintroduce organizational complexity through such devices as "agglomeration economies" to explain the persistence of cities, the "geography of enterprise" to capture the role of large corporations, and "systems analysis" to cope with the multitude of interwoven connections among plants, firms and cities. But they have never resolved the problem of organization. To do so, it is necessary to pause and question the categories themselves.

In this chapter we argue that organization and geographical industrialization should be considered in terms of a unified approach to the spatial division of social labor and modes of production integration. This means taking a fresh look at the nature of industrial organization itself, to see the organizational problem as an open one, with many possible answers. The units of modern industry are constructed as a part of the industrialization process, evolving as capitalists wrestle with the division and integration of production as an essential problem in production and competition.

The integration of production systems takes place through market exchange, within the firm, at the workplace, in the industrial sector, and in the territorial complex. These modes of organization treat different facets of the division of social labor and no one alone can solve the vast problem of integration (Kafkalas, 1985, Walker, 1988c). The result is a rich tapestry of organizational forms that link up and overlap in complex and flexible ways. In our view, industries and territorial complexes are the dominant modes of organizing production, not giant corporations. This goes hard against the prevailing wisdom which has recently jelled around the notion of a new spatial division of labor (Froebel et al., 1977; Dicken, 1986; Smith and Feagin, 1987), as well as the behavioral approach known as the geography of enterprise (Krumme, 1969; Hamilton, 1974; Taylor, 1975). In these views, there has been a decisive change from an age of manufacturing centers, such as coal and steel at Tyneside or textiles in Manchester, to the contemporary pattern of occupationally-differentiated places, such as skilled mental labor performed in London and unskilled manual work done in branch plants in southeastern Wales, organized by large firms (Massey, 1984; Cooke, 1986). The new spatial order is, furthermore, usually seen as arranged hierarchically in space according to corporate organization charts (Hymer, 1972; Cohen, 1981). We contend that industries are organized in such a way that territorial centers still play a vital role, despite the enlarged role of the corporate form of organization and the shifting nature of the social division of labor.

5.1 The organization of production

5.1.1 The problem of defining industries

The term "industry" is extraordinarily slippery. A factory is manifestly a thing, a firm is a legal entity; but what existence does an industry have? An industry is a congeries of firms and parts of firms (plants, divisions) embracing several related production activities. But because production systems are deeply intertwined, the boundaries between industries cannot be drawn sharply, nor is there a clear measure of the strength of relations within each industry. As a result, "What economists, management experts and others think of as industries often are merely collections of products and services that bear some relationship to one another" (Sobel, 1983, p. 139).

Industries might be defined in terms of final outputs: steel, cans, paint, etc. But every conventional output category can be broken down into literally thousands of different commodities, some with only generic similarities. The construction industry, for example, builds homes, infrastructure, office buildings, factories, etc., and each of these can be broken down further: single-family homes, apartment buildings, town-houses; freeways, suspension bridges, small roads, and so forth. These often involve different companies with distinct skills. Another possible criterion is the nature of the production process, such as machining. But one cannot easily characterize the diverse methods of production when complex processes are broken into many discrete parts. Telephone manufacture, for example, may include hand assembly lines, highly mechanized wiring and soldering processes, hand testing of components, semi-automated clean room circuit board production, plastic moulding, and metal parts machinery, not to mention clerical, engineering and maintenance work. In other words, production is always parcelled into several discrete tasks and work groups called the division of social labor.

The division of labor was a fundamental category of classical political economy, as in Adam Smith's parable of the pin factory, yet in this century it has been confined to a handful of theoretical explorations (e.g., Young, 1928; Coase, 1937; Stigler, 1951). The result of this neglect is a remarkable tendency to think in terms carried over from the days of simple commodity production, of manufacture and the Industrial Revolution. The classical distinction was between the social and the detail divisions of labor, the former referring to commodities (wagons, corn, shoes, etc.) and the latter to tasks carried out by individual workers within the factory (cutting, sharpening, testing, etc.) (Marx, 1867). The combined result of detail work was thought to be a final product ready to sell. But this cannot be sustained because, as input–output analysis illustrates, every production process connects to virtually every other within the general matrix of production.

The US Census Bureau's Standard Industrial Classification (SIC) system illustrates the definitional morass. SIC codes begin as a broad grouping of commodities and then breaks down into progressively more similar groups of outputs (e.g., 3 manufacturing, 35 machinery, 354 metal working machinery, 3541 machine tools, cutting). The census-taker immediately runs into the problem of collecting data by firm and classifying all employees or output by the major activities of each company. Worse yet, the final commodity basis of classification breaks down as intermediate goods appear. As a result, some SIC codes appear to center on end use (e.g., "transportation equipment"), others on material base ("non-ferrous metals"), still others on method of processing ("chemicals"), and yet others on component parts ("aircraft engines" and "airframes"). Furthermore, common-sense industry categories pop up at different levels – 3-digit, 4-digit, even 6-digit (Shepherd, 1979). In short, these conventional definitions are of very limited analytical usefulness (Kraushaar and Feldman, 1988).[1]

Input–output analysis ought to have placed the division of labor back at the center of economics, and forced us to recognize the inadequacy of

[1] A sample breakdown of SIC codes, to illustrate the *ad hoc* categories:

35 Machinery
351 Engines and Turbines
353 Construction Equipment
3532 Mining Machinery
3533 Oil Field Equipment
3534 Elevators and Escalators
3535 Conveyors
354 Metal Working Machinery
3541 Machine Tools, Cutting
3544 Specialty Dies
3546 Power Hand Tools
3547 Rolling Mill Machinery

existing definitional practices in industrial organization theory. In practice, there is no workable *a priori* definition of industry boundaries, since in any economy with a complex social division of labor it is impossible to assign activities to unique organizational locations. The lines between input and output, between commodity and non-commodity, workplace and firm, individual and collective labor are remarkably flexible. Therein lies our problem. In what follows we focus on the formation and integration of industries, on their dissolution and division; we regard industries not as fixed phenomena, but as moments in a dynamic process of division and integration of labor.

5.1.2 Workplaces, firms, and economies of scope

To solve the problem of how industries are constituted, we begin with the analysis of smaller organizational units, the workplace and the firm, before returning to the question of organization at the industry level. Here the literature on economies of scope and transaction costs provides some guidance.

The workplace is, first, the site at which labor and means of production are brought together; as noted in chapter 3, it is a preeminently geographical unit. Too often its geographical and organizational aspects are taken to be unproblematic, however. Traditionally, three kinds of workplaces have been distinguished: the home; the workshop; and the factory.[2] The great majority of manufacturing is done in factories, as is an increasing percentage of non-manufacturing work ("office factories"). The Industrial Revolution of the late eighteenth century brought the rise of the factory proper, but the factory did not sweep all of industry before it; throughout the nineteenth century, much of Britain remained a nation of workshops, as did the United States (Montgomery, 1967; Samuel, 1977). Factory production became much more widely generalized in the twentieth century (Nelson, 1975), yet, millions of small workplaces are still embedded in the industrial matrix of modern capitalism, along with store-front offices and retail boutiques. Home work was thought to have died out in advanced capitalist countries, but remains widespread in sectors such as garments, electronics, and insurance (Rainnie, 1984; Siegal and Markoff, 1985; Baran, 1986). In fact, the average employment-per-factory has been shrinking in recent years, in some cases spectacularly, as with the dismantling of the great motion picture studios (Storper and

[2] These are all closed workplaces. The other great realm, open workplaces, is rarely touched upon. Where labor is not bound to a fixed venue, there may be mobile workplaces, such as airplanes or ships; shifting workplaces, such as the sites occupied by construction, lumbering and on-location filming; and extensive work sites, such as farms and national forests. Some work is completely nomadic, such as traveling sales and photocopier repair. Infrastructural industries are characterized by systems of nodal workunits, such as railyards or powerplants, embedded in fixed distribution networks.

Christopherson, 1987). In this and many other industries today, the scale of most workplaces falls between the large Fordist factory and the small workshop: it is a large batch production unit occupying a single shell building, with a moderate internal division of labor. There is no common term for it – perhaps "batch factory" will do. The point is that widely varied activities are carried on within single workplaces, and organization theory has been unduly captivated by the imagery of Leviathans such as Henry Ford's River Rouge plant, employing 35,000, which took in coal and iron at one end and put out automobiles at the other. Bringing so many different labor processes and workers together under one roof is a major accomplishment, but it is the exception, because it presents so many organizational headaches. As a result small- and medium-size workplaces remain critical to modern production.

The firm is the other basic unit of production under capitalism. It has evolved as a legal shell under which assets can be assembled, contracts drawn up, workers employed, and capital accumulated. The modern firm is, first, an entity with which other capitalists can do business: it has an ongoing existence, legal rights and obligations, an address, etc. The firm is a container for capital in its various forms. Machinery, inventories, and buildings can be brought together or spread across the face of the earth without being assigned to particular people. Money capital can be accumulated in corporate accounts. The firm is an employer with the right to hire and fire and set the terms of labor within its bounds, a legally circumscribed place wherein capitalists may exercise their right to exploit labor-power. The firm is an administrative structure created to organize production, handle external exchanges, manage assets, and exercise class control. Finally, the firm is the central actor in the competitive battle among capitalists.

The scope of production activities encompassed by single firms is exceedingly diverse. Some small, single-plant firms fill specialized product or task niches within an industry, while giant corporations may span several industries and have facilities all around the world (Marris and Wood, 1971). Most small firms survive by doing things large companies do not want to do or cannot be bothered with: serving local markets, making highly specialized products, exploiting marginal laborforces. Even large firms usually have divisions doing specialized work. In the semiconductor industry, for example, there is a wide variety of chip technologies and product niches; as a consequence, a complex patchwork of firm strategies and sizes has emerged (Truel, 1980a; Schoenberger, 1986). Component-assembly production systems are also a haven for small firms. The typical airplane engine has 10,000 parts, mostly made in small batches by different firms (Storper, 1982). The chemicals industry includes innumerable downstream plastic fabricators, pesticide formulators, and the like (Commoner, 1976).

Why does this mix of Lilliputians and Leviathans persist? Or, as Coase (1937) put it, why is all production not incorporated within a single gigantic firm or distributed among millions of tiny firms? The real situation falls somewhere in the middle. That is, different production systems have characteristic degrees of organizational integration and disintegration. Put another way, production units (workplaces or firms) have greater or lesser range, or scope, of activities. The problem of scope has been taken up vigorously in recent years by Oliver Williamson (1975) and others of the "transactions costs" school, and has been brought into geography through the illuminating work of Scott (1983, 1988a). This model is usually posed in terms of firms, but actually applies to units of production in general.

Basically, under this model, firms have a choice between bringing a production activity within the bounds of the workplace or firm (internalization) or purchasing the necessary output from another workplace or firm (externalization). The goal of cost minimization guides this choice. Economies of scale, it is argued, are greatly overrated as a cause of firm expansion. True economies of scale refer only to cases of technically unitary equipment that yields increasing returns in terms of physical size and scale of operation, such as open-hearth furnaces; they also occur in work on large integral products, such as ships or airframes.[3] In most cases, however, the unity of product or equipment is not so clear-cut, and the size of the firm or workplace turns on advantages of joint operation of potentially divisible labor processes and machinery, or internal "economies of scope."

Internal economies of scope are present where it is more efficient to operate two or more activities in tandem than in isolation (Coase, 1937; Panzar and Willig, 1977, 1981). They are of several kinds:

- *Technical indivisibilities* involve direct physical connections of machines and materials, as in moving belt assembly lines, heat transfer between two processes, liquid flow through a contained system, or different by-products derived from a common resource (Bain, 1956; Bailey and Friedlander, 1982).[4]
- *Concurrent scale economies* occur where related production processes reach their least cost output at roughly the same level. Otherwise, joining them together will mean running one at suboptimal capacity.
- *Coordinative economies* are those where integration of labor processes and regulation of material flows benefit overall production, such as linking research and development closely with manufacturing, marketing a line of products jointly, or realizing efficiencies of continuous machine operation by providing a regular supply of inputs.

[3] For our somewhat different treatment of economies of scale, see section 2.3.

[4] The distinction between these and strict technological indivisibilities (one machine) can obviously be very fine in some cases.

- *The sharing of technical know-how and working skills* from one process or product to another is the most subtle form of internal economy of scope, but perhaps the most important (Caves, 1982). For example, potato chip makers have all moved into other lines of snack foods not only because a broader product line is desirable from the sales perspective, but also because they appear to have firm-specific knowledge that transfers from one product line to another.
- *Social control economies of scope* occur because the workplace and company property are places of containment and confinement, pieces of turf where the boss rules, and hence central to labor control (Marglin, 1974).

Conversely, internal diseconomies of scope encourage splitting tasks into separate production units. For example, if two related production processes have very different levels of least cost output, it may not make sense for a firm to produce the one with the higher minimum scale. Sometimes firms cannot master the knowledge required by different production processes and can buy better products from specialist firms. Large factories have political drawbacks as a large number of workers gather in a single place where they can develop a strong sense of their collective power. Since at least the time of the Flint sit-down strikes in 1937, some industrialists have intentionally moved to split up factories, second-source, and subcontract to evade organized labor (Bluestone and Harrison, 1982).

Internal economies and diseconomies of scale and scope alone are insufficient to determine whether related activities should be brought within the bounds of the firm or left to the market, however. This requires the additional consideration of transactions costs (Williamson, 1975; Panzar and Willig, 1977; Teece, 1985). Almost any production process can be divided into smaller parts, it is argued, and transactions can take the form of open market sales or be governed by any of a variety of contractual arrangements. Just because wool and mutton are joint products of sheep does not mean that a shepherd cannot contract out the shearing or slaughter. This model provides a powerful set of reasons why large size is not always the most efficient, and why small firms can survive by suitable marketing arrangements. Conversely, both markets and contracts operate poorly under certain circumstances and direct administration within a firm can be an efficient method of integration because the firm can command the necessary information and enforce the required behavior. The costs of administration depend, of course, on organizational methods and technologies, such as effective bureaucratic procedures or internal cost accounting.

The balance between economies of scope and transactions costs yields a characteristic degree of integration/disintegration in every industry. The concept of scope provides an entrée into the question of how individual units of production are defined and how this definition is itself influenced by a set of dynamic technological and political forces.

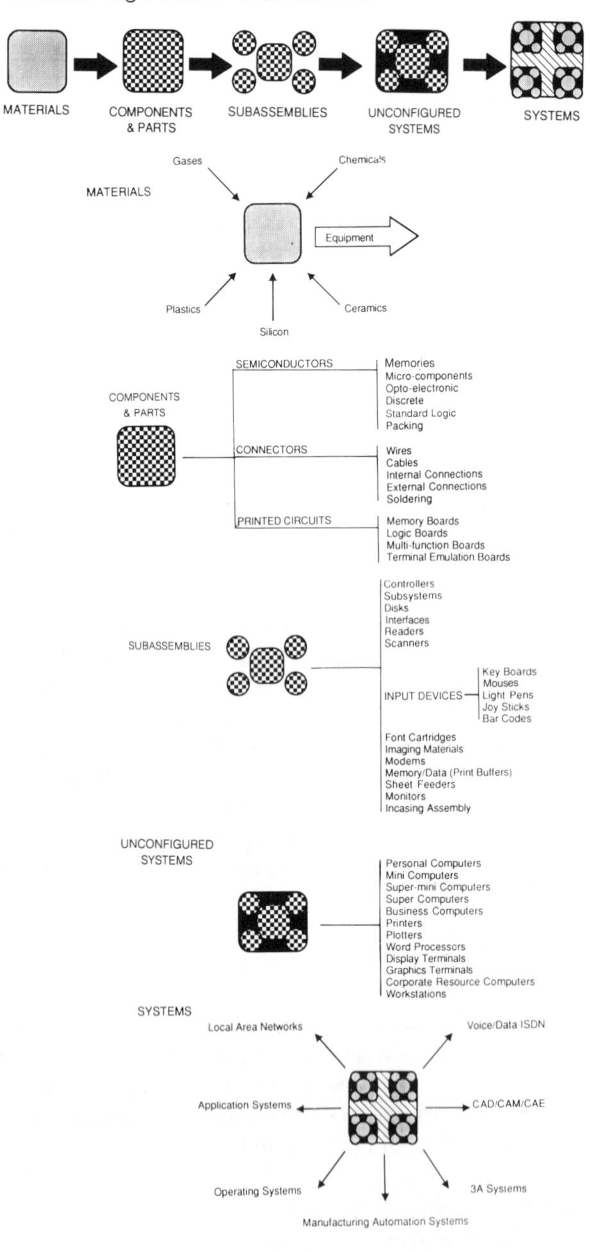

Figure 5.1 *Elements of a hypothetical "information industry" taking shape in the 1980s* This extended series of vertically- and horizontally-linked products, grouped around semiconductors and computers, is thought by some to define an emergent industrial sector or ensemble. Such schemas demonstrate both the web of relations among industrial activities and the difficulty of defining "industries," especially in technologically dynamic areas. (Based on information from Dataquest, Inc., San Jose, California)

5.1.3 Industries and production systems

The economy can thus be divided into hundreds of thousands of production units and each unit may take inputs from many places and send outputs in many other directions. In the end, the entirety of the diversified industrial economy is one big production system organized around that economy's division of social labor. This insight, while important, is most unsatisfactory for the purposes of industrial analysis. Some connections, after all, must matter more than others. There are technological relations that lend order to what would otherwise be a random network. It is possible to discern certain shapes or connections in the thicket of input–output relations, and these form an essential basis for the constitution of industries or of larger industrial ensembles. As a result, empirical studies of intra-industrial linkage show a fair degree of convergence on the definition of major sectors and ensembles despite an unavoidable measure of blurring and disagreement on certain particulars (Czamanski and Ablas, 1979). Production systems typically rest on a base technology, such as microelectronics (see chapter 4). The French term "filière" captures the idea of a connecting filament among technologically related activities (Truel, 1980a, 1983) (see figure 5.1).[5]

When do the ties bind closely enough to make the fates of production units strongly interdependent? Relations of necessity are central to the problem, but these are not determined in any simple way by technical relations: the steam engine does not give us the factory. For example, steel is an input into automobile production, and much of steel capacity depends upon demand derived from automobile sales. Yet, steel remains decidedly a separate industry from the car industry in the United States, despite Ford's and Kaiser's efforts at complete vertical integration. Why? To begin with, the input–output relations between steel and autos are only in one direction, i.e., autos must consume steel but the two sets of activities are otherwise independent, because steel is consumed by many other industries and the technologies of production have little in common. Parts of the steel industry form units in the production system for cars, but are not part of the car industry because they are strongly integrated into giant steel mills producing several types of steel products with complete autonomy from automobile firms. This brings us back to the concept of economies of scope and of the transactional structure of production. Steel production is not physically integrated into car factories today because there are internal diseconomies of scope to so doing, whether technological (scale), managerial (skill, know-how), or exploitative (separated labor markets) in nature. This does not entirely explain the border between the two industries, however, for it might still be possible for a car firm to own steel plants to produce output

[5] But the geography of production ensembles has not been taken in a fruitful direction in the French literature, in our view (e.g., Truel, 1980b).

for use within the car industry with excess output sold to other industries. Yet car and steel production today are not only physically disintegrated, but also legally and organizationally disintegrated; the borders of firms are drawn so as to exclude steel-making from the automobile industry. Apparently, hypothetical car and steel firms would face considerable organizational costs with few, if any, compensating advantages in terms of economies of scope; excessive integration can create untold management difficulties and technological confusion.

No single organizational solution will suffice for a given industry, however, and industries always consist of complex networks of small and large firms, big and little workplaces (whose boundaries do not always coincide with the industry's boundaries). The diversity of companies active in the US computer industry, circa 1987, offers a good example. IBM dominated the industry with a full line of products, but held its strongest position in mainframes; Honeywell and Unisys, among others, challenged it in this area. Digital Equipment Corporation held the pivotal position in minicomputers, thanks to its lead in scientific applications through the VAX system; Wang, Control Data and Hewlett–Packard were also active in this realm. Cray set the pace in supercomputers, joined by Amdahl. In microcomputers, Apple's Macintosh was the current favorite but many more, such as Leading Edge and Compaq, had made their fortunes by cloning cheaper versions of the IBM PC series. Sun Microsystems was moving rapidly to close the gap between micro- and mini-scale machines with its powerful line of personal computers. In short, because the problem of dividing and integrating social labor is so intractable, there will invariably be several forms of integration at work simultaneously in any industry.

National variations are readily apparent in the way every industry is organized; different solutions evolve to handle roughly the same problems. For example, the US semiconductor industry rests more on a small firm-venture capital mix than its counterparts in Japan or France, where semiconductors are mainly a branch of electronics giants, aided by state planning and ownership. There is no reason to think that the US model is necessarily the best choice, however, despite confident proclamations about the virtues of the entrepreneurial small firm (e.g., Zysman, 1977; Rogers and Larson, 1984).

What binds the diverse particles of economic activity in an industry together? The usual answer is markets. And indeed markets are powerful institutions and instruments of integration. Markets are distinguished by relatively individualized, voluntaristic, formally equal, and temporally constrained transactions between parties, they allow remarkable flexibility through a changing mix of participants and commodities, and sharp mobilization of personal energies via competition and the pursuit of self-interest. Nonetheless, markets are beset by difficulties. The principal limits to the formation of stable, workable market exchanges are uncertainty (incomplete information, an uncertain future), small numbers (few parties, irregular transactions), bounded rationality (inability to handle all information and contingencies), and opportunism (misrepresentation, reneging) (Alchian and Demsetz, 1972). As a result, commodities may not be available as needed, critical information may be withheld, and labor processes may be poorly coordinated. Take the case of unequal or impacted information in a relationship between producers who control different production units (Williamson, 1975). One party to the transaction may be able systematically to expose the other to a loss of income. For example, one firm producing inputs for another might profit opportunistically by using inferior materials if the consuming firm cannot easily monitor the quality of the input . Some measure of vertical integration may result, where the consuming firm purchases the input-producing unit. More frequently, however, this type of fundamental interdependence comes to be governed by something other than a spot market.

Industry governance systems come in several forms: subcontracting, strategic alliances, trade organizations, state planning agencies, or informal relations of trust between firms in a business community. The growth of large firms in the twentieth century has virtually blinded researchers to the importance of these interfirm methods of governance. The conventional analytical dichotomy between firm and market elides the way the world outside the firm needs to be managed and the way the world inside the firm needs to be regulated in conformance with external relationships.[6]

A particularly prominent form of intra-industry governance is subcontracting. Subcontracting offers a compromise between open-market transactions and intra-firm internalization (Belil, 1985; Holmes, 1986).[7] It implies stronger linkages, coordination, and regulation than the open market, but is still formally defined by contractual rather than authority relations. The subcontractor is legally independent (unlike the subsidiary) and free to undertake other contracts (unlike the regular employee), though lead firms may direct subcontractors by providing materials and equipment, specifying product and process, presenting designs, licensing patents, training personnel, monitoring production and management, loaning specialized workers, etc. Subcontracting overcomes certain deficiencies of the open market (e.g., uncertainty, fragmentation) and certain problems of the firm (e.g., labor relations, managerial overhead).

[6] Indeed, even inside big firms, management has introduced such devices as divisional organization, profit centers, and *ad hoc* work teams to try to emulate market conditions of profit equalization, competition and small firm coordination (Chandler, 1962, 1977).

[7] Putting-out is the oldest specifically capitalist form of organizing production, and predominated in many parts of Europe in the eighteenth and nineteenth centuries (Kriedte et al., 1977). Thought by many to have passed from the stage of history, it is nevertheless to be found today in automobiles (Holmes, 1986), garments (Scott, 1988a, ch. 6), films (Storper and Christopherson, 1987), computers (Bakis, 1977), agriculture (Fitzsimmons, 1986), insurance (Baran, 1986), semiconductors (Dosi, 1984), aerospace (Scott, 1988a, ch. 9), machine tools (Brusco and Sabel, 1983), and so on.

Another prominent form of intra-industry governance is strategic alliance. Firms have tried to stave off competition, combine proprietary technology and know-how, market a wider variety of goods, and so forth through such strategies as merger, overlapping ownership, interlocking directorates, cross-licensing and joint ventures (Harrigan, 1985). An industry-wide alliance of note is the US semiconductor industry's new joint venture, Sematech. Still another form of external industry management is the trade association, which arose as an experiment of the late nineteenth century (Galambos, 1966). These have little direct power over companies, but allow an industry to present a united political front, undertake joint research projects, provide information and services to their members, and engage in industry-wide labor recruitment, wage bargaining, and strike management. The state has also entered the picture on occasion. In the United States, for example, the War Industries Board of World War I, the National Recovery Act of 1933–4 and the Office of Price of Administration in World War II were powerful methods of regulating competition and production in periods of emergency. Today, industry-wide or regional consortia are working to develop new technologies, as exemplified by the European Community's Esprit and Eureka projects in electronics, or the Midwest states' Advanced Ceramics and Composites Partnership. Japan's method of intra-industry and state regulation are presently much admired, especially the role of the Ministry of Trade and Industry (MITI) (Johnson, 1982).

Thus, we can define industries as particular organizational groupings built around extensive production systems, in contrast to the more restricted compass of the workplace or firm. Industries embrace relatively large numbers of related production activities whose interdependence demands a measure of material linkage and labor process coordination higher than that ordinarily achieved through open market transactions alone. That is, we hold that intra-industrial transactional relationships are so qualitatively distinct from inter-industrial transactions that they create a meaningful definitional boundary for the industry. Within industries, the division of labor will be integrated through a combination of different size workplaces, specialized and generalized firms, and a fabric of interfirm governance; by contrast, relations between industries will be largely regulated by open market relations. No industry will consist of only one type of production unit, or only one type of governance mechanism. Industries are comprised of heterogeneous production units and coordinated through diverse forms of production integration and governance.

5.1.4 The dynamics of industry borders

The dynamics of industry creation and redefinition over time must be situated within our theory of economic growth. New divisions of labor in production continually give rise to distinct productive tasks and production units. This is not due simply to the appearance of new products alone, but also to new ways of organizing productive activity. Occasionally, groups of activities arise that have particularly dense input–output relations, or are grounded in a distinctive technological framework. As such new realms emerge in the social division of labor, they may become knit into the coordinated systems of social and technical practices called "industries" (see figure 5.1).

Industries develop as self-governed systems of production. The industry is institutionally self-defined, where the division of social labor sets the possibilities and limits of the modes of integration and governance system. This is an exploratory process of adapting to new conditions and inventing new methods of integration and governance. In this process, neither the definition of industries nor their organizational fabric is based on short-run minimization of transactions costs alone. Both are constructed actively through the process of strong competition.[8]

Take, for example, the rise of the office-based industries, which in many ways parallels the earlier separation of manufacturing from household production. New commodity markets have appeared for what used to be intermediate goods and labor-services provided within the manufacturing corporation (Walker, 1985a). Everything from advertising copy to software, from product design to legal advice, is now frequently secured through market or quasi-market transactions. The production of these "business services" has now achieved a level of autonomy and internal complexity on a par with any other industry. Many manufacturing industries have a similar history: for example, when aluminum was a specialty by-product of the chemical industry, no one was concerned with the organization or location of the "aluminum industry". What was once produced as part of the internal division of social labor within existing

[8] Competitive advantage may result from attempts to redistribute income between different producers (as when large firms exert price pressure against vulnerable supplier firms), attempts to assign risk (in which case minimizing transactions costs is subjacent to minimizing long-term production costs), attempts to encourage other firms to innovate while securing the benefits of that innovation (long-term strategic alliances), and so on. Efforts to minimize short-term transactions costs do not guide capitalist behavior in developing governance mechanisms (and hence in defining industrial sectors). The fundamental insights of Coase and Williamson must be set within more realistic theories of technological development and competitive behavior to generate a more robust view of the dynamics of industry division and redivision. We ignore, for the moment, labor and employment relations: organizational dynamics flow, in part, from efforts to divide workers, control communities, and decrease labor's share of income (see chapter 6); but, in our view, organizational change is not principally induced by labor struggles.

industrial sectors has now come to be an external division of labor between industries.

5.2 Territorial production complexes

5.2.1 Territory and the organization of production: a reconsideration

Cities have an immediate material presence for everyone, much more so than General Motors, and nations bear on our lives through wars and taxes in a way Westinghouse can only envy. In other words, we have no problem thinking in terms of territory in everyday life or for that matter letting it slip into academic inquiry, with such geographical monikers as "Route 128," "Wall Street," or "The Sunbelt." The problem, then, is theorizing the territorial organization of industrialization in a cogent way. Territorial formations such as cities must be understood to involve social dynamics set in motion as a consequence of dense human settlement (Mumford, 1961; Lefebvre, 1974). These interactions between economy, geography and society have largely been omitted in work on industrial location.

The prevailing approach in industrial geography regards location patterns as the result of prior organizational forms and decisions. Weberian theory works out the calculus of single plant siting choices; enterprise theory looks for the intra-firm logic of giant corporations; oligopoly theory argues that the behavior of a few large firms determines location. Against these views we hold that territorial production complexes stand on their own as modes of industrial organization that contribute powerfully to the dynamics of industrialization.

The territorial complex is an extensive work site that brings disparate production activities into advantageous relation with each other, at a larger scale and scope than the individual workplace, firm, or even, in many cases, the industry. Territorial formations are fundamental to the operation of industry because they offer means of integrating production systems above and beyond those found in organization charts, market institutions, or the laws of ownership and contract. Those means are essentially geographical in nature. Three principal aspects of spatial interaction are at work here: simple propinquity minimizes the costs and effort of movement, improves access and pools resources; locational fixity of infrastructure and daily activity establishes a resource base, lowers uncertainty and information costs, channels movement and reduces social distance; geographical boundaries limit movement, turn social interaction inward and solidify (and differentiate) social relations. In very general terms, territorial complexes not only lower tangible costs of transport and communication, but ease information-sharing, allow pooling of labor and fixed capital, stabilize physical and social relations, help people identify with each other (and against outside competitors), and generate distinct cultural practices over time.

To be sure, territories are not associated with unitary agents: they are extremely loose, open and non-hierarchical modes of production organization compared to the factory or the firm. This makes their functions difficult to visualize for those not accustomed to seeing things in geographical terms. They have much in common with markets: the institutional fabric of both is very thinly drawn; both rely heavily on voluntary interaction with few formal rules or administrative apparatuses; and both allow extremely flexible relations among many parties.[9] It is precisely such attributes of these complex socio-productive systems that make them so pervasive a feature of industrialization and require our further analytical attention.

5.2.2 The city as a territorial mode of production organization

Cities are dense clusters of production activities, people and infrastructure (the built environment). Urbanization and industrialization are inextricably linked through the process of agglomerated industrial growth (Scott, 1986b). Cities allow a degree of integration of production with a minimum of central control and a maximum of flexibility; in that sense, they are a highly suitable form of social organization for the anarchic side of capitalism. The development of the division of social labor necessarily implies some level of vertical and horizontal disintegration of production. Industries, as we have noted, are essentially groups of production activities held together by some form of governance system and almost always involve multiple production units and many firms. Those relationships can have geographically-sensitive cost structures, particularly where transactional relationships between production units are especially dense. The greater the costs per transaction, the greater the probability that firms will agglomerate in order to reduce them. Three types of transactions are especially affected by distance: those that cannot be standardized – that is, are unforeseeable – and require frequent search and recognition (these appear where markets and product designs change frequently); small-scale linkages which cannot enjoy volume discounts on transport costs; and problematic linkages that must be resolved through personal contacts or renegotiation. All these types of transactions are common in capitalist production systems, hence spatially-concentrated complexes of economic activity are very likely to appear on the landscape of capitalism. These agglomerated industrial complexes, as we pointed out in chapter 3, are

[9] Of course, at every territorial scale – city, region or country – institutions of governance, such as local business councils, regional planning agencies, or ministries of trade and industry, are also formed (Heiman, 1988).

the geographical means by which firms realize external economies of scale in production systems. That is, the inherently social nature of capitalist production is frequently realized in territorial forms of production organization.

Urban concentration benefits producers in four main dimensions. The city improves exchange by making comparison easier, by pooling diverse buyers, sellers, and information, and by letting buyers and sellers get to know and trust each other personally, and complete specialized or extended transactions. It is where you go to be on top of the action, to find the merchants, brokers and others who are in the know about market conditions, which is especially valuable in uncertain or rapidly-changing markets. Spatial concentration aids inter-workplace integration by facilitating worker movement between sites, managerial oversight, information transfer, evaluation and feedback across disparate parts of related production activities. The city keeps firms on top of technological know-how by letting them stay abreast of the latest information, draw on the most knowledgeable workers and managers, work on the specific problems of customers, and find the most diverse and creative suppliers of components and solutions. Cities are, finally, great labor markets for workers of the most varied skills, whirlpools of humanity drawing immigrants from rural areas and from distant lands to replenish the stock of cheap, pliable, and diligent workers; they also tend to draw too much labor and keep a surplus available.

As a result, industries have repeatedly (and in some cases continually) dominated national and even world trade from a base in a single city: silk in Lyon, cutlery in Sheffield, guns in Birmingham, securities in New York, commodity trading in Chicago, mining equipment in San Francisco, shoes in Lynn, or raisins in Fresno (Sabel and Zeitlin, 1985). These complexes all produce a wide variety of specialized intermediate and final outputs due to the flexibility of their division of social labor. The extraordinary myopia about the quantity of production in big cities such as Paris and New York reflects the fact that so much of it is hidden away in small workshops and firms that have not been brought under the aegis of the factory and the large corporation. But cities are not exclusively the home of small workplaces, nor do big factories invariably move out of cities. Many of the largest cities are dominated by large-scale workplaces: Pittsburgh, Detroit, and Seattle, for example. Even in areas that otherwise exemplify the small firm agglomeration, one finds big factories – Lockheed and IBM facilities employ roughly 20,000 and 15,000 people in Silicon Valley, and the Hughes Aircraft plant employs more than 15,000 in Orange County – and these large industrial plants are often embedded in industrial systems with intricate inter-linkages between many production units. They, too, profit from external economies of scale.

The force of spatial agglomeration can be seen in the way big cities and regional complexes draw distant industries and plants, "distorting" the ordinary tendencies to localization and dispersal among industries. Agglomeration makes any plant-by-plant optimization of location impossible, as even Weber (1909) acknowledged. Clustering is foremost among the industry-level patterns of geographical industrialization.[10]

5.2.3 Large-scale territorial development: city systems and regions

Territorial production complexes can take a variety of spatial forms. One is the large city or metropolis, such as greater London or Baltimore, which contains a number of industries and their specialized districts (Hall, 1962; Muller and Groves, 1979); a perfect example is Philadelphia's turn of the century textile matrix: knit goods in Germantown, carpets and spinning in Manayunk, underwear and seamless hose in the suburbs (Scranton, 1983). Another is the city-satellite system, in which a number of small towns are tied in essential ways to the dominant metropolis, as with the textile towns scattered around Boston or Manchester in the nineteenth century, or the mining towns of the Mother Lode and Nevada centered on San Francisco after 1849 (Pomeroy, 1965; Vance, 1977). A third form is the cluster of towns, in which none is overwhelmingly dominant, as in the textile region of North Carolina or the metalworking towns of the Connecticut Valley where jet engines are produced (Storper, 1982). The fourth and largest regional complex is the manufacturing belt of the kind that developed across the American Northeast, the West German cities of the Rhine and Ruhr, or the Osaka–Nagoya–Tokyo belt of southern Japan. Such territories are so immense that they have been virtually identified with national or continental industrialization as a whole (DeGeer, 1927; DeVries, 1984). It is striking how territorially concentrated industry remains when looked at from a global scale, despite two hundred years of capitalist expansion.

Large scale forms of territorial production organization enjoy the same types of external economies of scale as would the single agglomerated sector. A polysectoral production region – whether organized in the form of a large metropolis, a city/satellite system, a cluster of towns, or a manufacturing belt – is like a spatially-extended agglomerative field based on a more elaborated division of social labor than would be found in a single sector in a dense urban industrial district. Nonetheless, the analytical foundations for understanding such fields are very much the same. As research on systems of cities has shown, extensive territorial complexes incorporate trade and activity networks through whose channels flow deep and swift currents of goods, labor, information, and money (Pred, 1977;

[10] This is rediscovered, it seems, once a generation: by De Geer (1927) as the "manufacturing belt," by Perroux (1950) as "growth poles," by Pred (1966) as "cumulative causation," and by Scott (1988a,b) as "disintegrated production complexes."

Bourne and Simmons, 1978). Major transportation and communication arteries cement these linkages, as do the filaments of personal knowledge, institutional ties and cultural practices. As a result, systems of cities, consisting of both the largest metropolitan areas and relatively small industrial towns, have long formed the backbone of regional and national development (Muller, 1977; Pred, 1980; Meyer, 1983). City systems are not just the products of inter-regional trade, however; they derive from intra- and inter-industry patterns of integration and agglomeration. Depending on the need for interaction among units of production systems, industries and ensembles of industries may form a single dense cluster (in a large metropolis) or a more scattered system of clusters (in a regional system of towns and cities). All such systems will tend to be multi-nodal as long as they involve more than one distinguishable industry. Considered at all possible levels of industry aggregation and specificity, the matrix of territorially concentrated industrialization produces a more or less hierarchical or uneven form of territorial development. Because there are so many differences in the mix of industries, organizational practices and national histories, it is not surprising that national patterns of territorial unevenness vary considerably.

5.2.4 A requiem for the geography of the corporation

The priority of the link between urbanization and industrialization disappeared from view for many years in favor of theories of a new spatial division of labor based on the corporation (Watts, 1980; Taylor and Thrift, 1982). Some location theorists went so far as to argue that agglomeration economies are completely inoperative in the corporate economy, on the grounds that material linkages, labor pools, and information are all internalized and commanded by the large firm (McDermott and Taylor, 1982). Others said that urban agglomerations of industry are held together by the oligopolistic behavior of large corporations (Markusen, 1985). Still others attributed the continuing vitality of cities to the retention of corporate headquarters (Cohen, 1981). In fact, industries and territorial complexes still prevail over firms as the building blocks of geographical industrialization.

The evidence for a distinct geography of enterprise, or corporate pattern of location, is surprisingly thin, despite all the research devoted to its verification (Hayter and Watts, 1983). While the force of large companies in the modern capitalist economy cannot be gainsaid, the division of labor is not simply a creation of the corporation nor is the firm the only means of productive integration; as a result, one must be cautious about attributing to the large enterprise locational effects better explained by other causes. The production conditions of industries still override the organizational impact of large firms in most cases. For example, in Britain it has been found that the employment loci of large and small firms by industry are so strongly overlapping as to be indistinguishable from each other, and that operating divisions of large companies manifest divergent locational patterns across industries (Watts, 1980).[11] Two case studies, of ICI (United Kingdom) and Phillips (Netherlands), which are regularly cited to prove the case for a distinctive geography of large companies, actually show just the opposite. The chief factors in ICI's massive geographic restructuring of the 1970s were product shifts, changing resource costs, market size and the search for new outlets, relative labor costs, and the nature of production at different plants (Clarke, 1985). Phillips' global moves have similarly been governed by a search for markets (chiefly government military contracts), cheap labor, advanced technological capability, and political stability (Teulings, 1984). In both cases the multinational corporation has been a facilitator of global operations through its ability to integrate a substantial number of large and technically sophisticated operations, but the parameters of their geographic production decisions are remarkably similar to those that limit smaller firms operating in the same industries.

In the same fashion, it has not been demonstrated that an intra-corporate organizational hierarchy is the principal source of locational differentiation among such activities as management, research and development, and manufacturing. It appears, rather, that each of these activities chooses a location with respect to its own position in the spatial division of labor, and this frequently means situating to take advantage of agglomeration economies and systems of territorial integration. Headquarters and research and development, for example, cluster in cities for ready access to business services, to each other, to skilled labor pools of managers and researchers, and so forth (Hoover and Vernon, 1959; Armstrong, 1972; Malecki, 1985). This is not a geography of the corporation so much as the geography of an extensive capitalist division of labor. That spatial division of labor predates the modern corporation since finance, trade and specialized business services have long clustered in large cities (Pred, 1966; Vance, 1977). On the other hand, the dispersal of manufacturing plants, as previously discussed, is motivated by such forces as standardization of external transactions in large factories, and opening up of new markets. This process, too, has a history before the corporation, as illustrated by textile mills established outside Manchester (UK) or Boston (US) early in the nineteenth century, where the principal capitalists were merchant companies.

The continuing pull of different forms of territorial integration, industrial production systems, particularly dense urban or regional agglomerations, on industrial production systems helps explain why predictions that modern

[11] The evidence does not show, contrary to prevailing opinion, that manufacturing plants of large firms have fewer local linkages than plants of small firms, except in the producer-services category (Lever, 1975; Martinelli, 1986).

capitalist societies will transcend urbanization invariably come to naught. Indeed, in the United States, after a brief period during the 1970s when non-metropolitan population growth rates exceeded those of metropolitan areas (Erickson and Leinbach, 1979), rapid metropolitanization has returned in the 1980s. Throughout the 1970s, moreover, metropolitan areas in the advanced countries continued to add greater absolute quantities of economic output, population, and income than non-metropolitan areas. It would appear that we have not reached an era of "counterurbanization."

5.3 Dynamics of territorial organization

We can now move beyond the rather static formulation of the problem of the division and integration of production to consider industrialization in motion. Here, we argue that the developmental trajectories of workplaces, firms, markets, and industries are decisively influenced by their embodiment in a territorial formation.

5.3.1 Territory and the dynamics of production

In a dynamic context, territorial clusters stand out as a very favorable mode of organization under the disequilibrium conditions of capitalist competition. In particular, territorial production complexes promote the flexible integration of industrial systems. Industrial systems do not simply embrace more advanced divisions of labor, but seek more fluid divisions of labor in which linkages, contracts, and production commitments can be continually made and unmade as an industry evolves. In a few times and places, these production networks have been unable to compete with emerging forms of mass production: in the mid-twentieth century, for example, large corporations were able to stabilize markets and product designs, and to institute technologically-unified systems of mass production that largely internalized the division of labor (Piore and Sabel, 1984). Nonetheless, over much of industrial history the outcome has been otherwise (Sabel and Zeitlin, 1985). We will emphasize three principal dimensions of production flexibility that are made possible by the formation of the territorial production complex: flexibility in the division of social labor, flexibility in the local labor market, and flexibility in technology. These are essential to the productive dynamism of complexes. That is, cities and industrial regions are not simply the geographical expressions of the growth of external economies in production; spatial concentrations themselves can also be a source of growth in industrial productivity. In short, territorial complexes alter industrial history.

Shifting techniques and unstable demand require flexible input–output relations and variable output levels. First, therefore, some degree of organizational disintegration arises frequently where production takes place under the combined conditions of growth and uncertainty.[12] Disintegrated systems are able to maintain economies of scale (in the form of external economies) through the territorial reintegration of social labor, and permit the industry to avoid reverting to low-productivity, artisanal production techniques in response to uncertainty and change. Once in place, inter-firm relations frequently have powerful effects on the further development of the division of labor. Territorial production networks both accommodate and encourage startup firms, experimental workplaces, and new work specializations. The very fluidity of territorial organization creates a grid of product and process opportunities, which provides the base for new producers as they go about the business of developing even more products or production methods (Scott and Storper, 1987).

Second, territorial production complexes permit the development of flexible labor markets, which, in turn, create new opportunities for production organization. The larger and more dense the local labor market, the more efficient will job-seeking tend to be for workers; similarly, employers in large local labor markets have a better chance of finding a job seeker with specified characteristics in a given time.[13] Accordingly, firms in large local labor markets can adopt flexible labor turnover policies and externalize their labor market relations to a relatively high degree; this allows them to respond more effectively to economic fluctuations and to engage in strong competition. Firms in small labor markets, by contrast, are more prone to hoard their workers because of the greater difficulties of finding replacements.

Third, the flexible milieu of the territorial complex promotes technological innovation because each center comprises a stock of know-how, labor skills and firm capabilities. Marshall (1900) recognized this contribution to the dynamics of capitalist development by noting that skills and information were "in the air" of industrial districts, such that the "mysteries" of production "become no mysteries" (see also Bellandi, 1986). In the process of conducting business in territorial networks, one learns about the manifold facets of production in the local complex. This leads to the

[12] In normal usage, uncertainty means chiefly unpredictability of markets (e.g., Berger and Piore, 1981; Williamson, 1975). But the catch-all use of uncertainty begs the question: what is the source of dislocation of exchange relations? Uncertainty is a universal characteristic of human life, but the particularly virulent form of uncertainty that afflicts industries is due to the disequilibrium nature of capitalist growth. That is, it lies in the constantly shifting basis of production owing to technical change in products, methods, and input–output linkages; the rising and falling fortunes of participants (workplaces, firms, cities, or industries) in the competitive struggle; and the ebb and flow of prosperity due to investment cycles.

[13] Jayet (1983) has formalized these relationships with the observation that workers in urban areas tend to alternate between employment and unemployment frequently, whereas workers in non-urban or small labor markets tend to experience more prolonged bouts of unemployment once they lose their jobs. Unemployment for urban workers, even in dynamic industries, can still be considerable, however (Shapira, 1986).

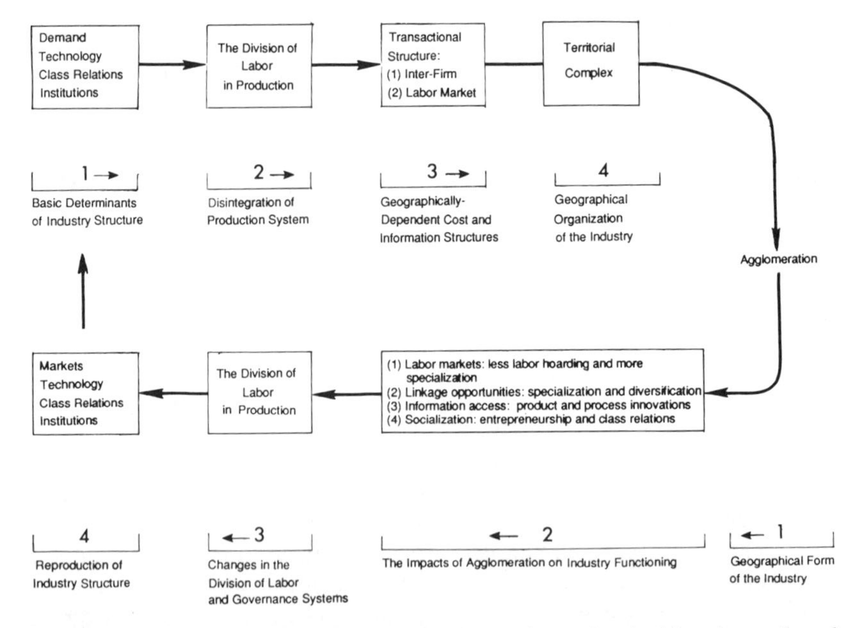

Figure 5.2 *Reciprocal influence of vertical disintegration and territorial agglomeration of an industry* In this schematic diagram, industry structure generates an agglomerated complex, which reproduces, in turn, that structure and its associated division of labor.

formation of a business culture in which practical forms of knowledge about production processes and markets are socialized, and tastes and sensibilities about materials, machines, and product designs are refined. For example, innovation in specialized tile equipment in the Sassuolo district of the Third Italy has been stimulated by the close interactions between tile producers (with their needs) and equipment producers (with their specialized capabilities) (Russo, 1986). In some cases, the socialization of useful knowledge is carried further by the educational infrastructure. The people involved in the daily life of each complex can most easily perceive new technological possibilities, and are also most quickly able to take advantage of opportunities, because the whole system implies high levels of turnover and mobility of personnel and firms (Scott and Storper, 1987). The development of sensibilities about product design can be observed equally among the creators of fine woolen fabrics in Italy, designers of specialized aerospace equipment in Los Angeles, or makers of small appliances in Germany. In all these cases, entrepreneurship and technical innovations are collective activities.[14]

[14] Territorial complexes can also become the sources of onerous rigidities: through land markets, the built environment, the politicization of local government and local labor markets, through entrepreneurial habits and attitudes, and so on. We shall have much more to say about these in the following chapters.

Figure 5.2 illustrates the two-way relationship between industrial organization and industrial location. The development of the industry's division of labor has powerful effects on whether it takes an agglomerated or deagglomerated form. The critical point here is that the geographical concentration of industrial systems permits those industries to function in ways that affect the developmental paths of the industries themselves.

5.3.2 Agglomeration and economic theory

It is germane at this point to consider how our view of territorial complexes goes beyond prevailing theories of urban agglomeration in marrying history and geography to capture the dynamics of industrialization. Neoclassical urban economics has difficulty explaining the build-up of large cities: in a pure neoclassical world of factor substitution and constant returns to scale, spatial concentration should never occur. Firms would adjust locationally as well as technically to the greater expense of factors in developed areas due to land costs, labor organizing, and so on. Indeed, the production system ought never get to the point where it would support large-scale workplaces, infrastructure or labor markets. Neoclassical theorists have gotten round this obstacle by introducing scale economics selectively into their models. Mills (1972), for example, assumes the existence of urban infrastructure with returns to scale (ports, railyards) and then deduces the growth of industrial agglomerations. Central place models assume increasing returns up to a certain scale but constant returns thereafter, hence activities sort themselves out in the urban system according to the market area threshold they require to exhaust their internal economies of scale (Losch, 1944). Still other models of "external economies" note that industrial activity enjoys higher productivity when spatially concentrated (Gaffney, 1962). They measure the costs of production at different urban scales and with different industry mixes, and then work out partial equilibrium solutions for spatially efficient resource allocation. In the end these exercises amount to calculating aggregate production functions for the city and its industries.

In short, there is no consideration of how economies of scale might be created. Agglomeration just occurs due to the pre-existing, exogenously-determined technological attributes of production. The relevant question is how places create locational advantage endogenously in a process of development. A more challenging view of spatial concentration is, therefore, the disequilibrium model of regional development posited by Myrdal (1957) and embellished for the urban case by Pred (1966). Recall that this theory combines several elements: agglomeration economies of the market threshold variety; migration of labor and capital from rural to urban areas; Keynesian income multipliers; and the concentration of technical innovation through information flows among large cities. Growth is propelled, in this view, chiefly by means of an advantageous circulation of factors, income, and information through the spatially concentrated urban system, but the dynamics of production – particularly technology, the division and integration of labor, and relations with workers – are largely ignored.

Clearly, territorial complexes generate comparative advantage by building up agglomeration economies. The division of social labor, and the development of the roundaboutness of production, is a principal source of these increasing returns. In this view of capitalist industrialization, scale effects and their organization into the division of social labor drive territorial concentration. Product proliferation and increases in productivity are the prime determinants of location, not cost minimization and factor prices. Even income multipliers and the circulation of information only amplify what already exists. But what matters most is the ability to produce what did not exist before – new products, new divisions of labor, new machines, new firms and factories, new surplus value. Conversely, location is central to industrialization not because it helps minimize spatial costs or reduces circulation time, but because territorial production organization creates milieux in which the creative aspects of the economic process are facilitated and driven by strong competition. Urbanization is not simply a set of transactions between a given set of producers and the regional environment, or a container for a set of production activities; industrialization and the urban environment develop each other endogenously, and can continue to keep doing so as long as the forces of production continue to develop.

5.3.3 The rise and fall of territorial-organizational complexes

Industrialization and urbanization need not support each other indefinitely, however. This is an important exception to the general theory of agglomeration. The organizational and competitive basis of territorial clustering may break down, as predicted by profit- and product-cycle models (Markusen, 1985). But it can also be reconstituted on a new basis, in the same or a different place. As industries go through rounds of production restructuring, so they also go through major organizational overhauls (Massey and Meegan, 1982). Just as dramatic change in products or processes frequently gives rise to a window of locational opportunity, so the sector often enjoys a window of organizational opportunity. Industries simultaneously undertake new product and process designs, adopt new forms of organization, and establish new growth centers and peripheries. In such extraordinary times, an industry does not relocate in response to "siting decisions" of a known group of firms, as conventional location theory would have it. Rather, the whole organizational ensemble is likely to change, bringing a whole new set of players onto the field: new factories, new company divisions, new plant sites, to be sure, but also entirely new firms, new cities and regions, new supplier linkages, new merchants and financiers, and even whole new industries (or redefined industry boundaries). Thus, when a portion of the motion picture industry broke off from the existing industry in New York and moved to Los Angeles in the early twentieth century, it quickly introduced such organizational innovations as the studio system and theater chains tied to the big producers (Storper and Christopherson, 1987). During World War II, the US War Department awarded greater than half of its ship building contracts to West Coast construction firms because they had achieved significant breakthroughs in mass organization methods and labor mobilization from their experience in building the first high dams at Boulder Canyon and Grand Coulee (Wollenberg, 1988).

The prevailing view of industrial organizational development is inadequate, in our view. It sees a natural evolution from small units to large, propelled by the imperatives of scale and technological advance (Chandler, 1962, 1977; Galbraith, 1967). It would be wrong, however, to think that only the giant corporation has enlarged its integrational capabilities, with other modes of production organization – markets, subcontracting, small firms and cities – falling farther and farther behind. Changes in the technology of transport, communication and organization have worked to the benefit of all these modes.[15] Indeed, many large corporations have recently been shrinking, as they hive off extraneous parts that they can not manage well. The consequence of these virtual across-the-board improvements in integrative capability is that the menu of organizational choices remains large as industrialization proceeds.[16]

A contending theory depicts epochal shifts from one territorial organizational form to another through industrial restructuring: craft workshops and urban-based industries gave way to Fordist factories and Sloanist corporations, which are in turn being ousted by flexible production systems (Piore and Sabel, 1984). We propose to explain this kind of organizational restructuring in terms of increasing returns and the general theory of disequilibrium growth, although we shall suggest that craft production and mass production are only two among many possible organizational forms that industrial production may take. Once set into place, any territorial-organizational framework of production has an inertia that sustains it in

[15] Market institutions have been steadily enhanced by such innovations as the telegraph, the railroad, steam packet, stock exchanges, commodity exchanges, investment banking, telephones, digital telecommunications, computerized data management, and so forth (Harvey, 1975; Pred, 1980). Subcontracting networks have never been more ubiquitous than today; hence the flowering of the "informal" economies and "Third Italies" of the world (Bagnasco, 1977; Portes and Walton, 1981). Industries have also become more tightly knit with the development of national and international markets, large firms that compete across traditional boundaries, factories and disintegrated complexes that can produce by joint ventures among national firms, and so forth.

[16] A variant of the evolutionary model is that every industry must recapitulate the general history of organizational expansion, moving from a youth of small firms to an old age of entrenched oligopolists (Vernon, 1966; Markusen, 1985; cf. Storper, 1985).

a self-reinforcing way. Institutions are established, rules and behavior take on the weight of tradition, and cumulative causation takes its course (Nelson and Winter, 1982). Of course, organizational ensembles do not exist in a vacuum, but dominant frameworks tend to persist until a major shock makes them unviable – the same kind of overshoot phenomenon we have noted with regard to investment and technological progress (see section 2.5). When these shocks come to be felt, producers in the industry are likely to begin experiments in alternative forms of organization designed to respond to changing competitive conditions.

In motion picture production, for example, the studio system was gradually dismantled after the industry's steady decline through the 1950s, owing chiefly to anti-trust action against studio-owned theater chains and the introduction of television. Several responses were tried, but the one that gained ascendancy was a highly disintegrated complex, involving a greater variety of film and video productions with most films made by independent production companies, more flexible contracts with writers, actors, producers, etc., and a flourishing of specialized shops for editing, special effects, recording, film processing and so forth. The big studios fostered many of these changes, and tried to retain control over them, but the whole structure of production shifted decisively toward disintegration. Studio back-lots were sold off to raise cash, on-location shooting increased for greater realism and to avoid union work rules, new technologies emerged to aid on-location work, and these put special effects houses in great demand, and new forms of labor market segmentation arose. Eventually the whole industry was transformed into a more diverse and supple complex with a wide variety of products (Storper, 1988); in the end, a new organizational ensemble stood on its own.[17] There is evidence that similar processes are now occurring across a wide range of new and old industries (Cusumano, 1985; Becattini, 1987; Storper and Scott, 1988).

Such developmental processes cannot be understood with conventional analysis of the choice of organizational techniques because multifaceted shifts in production techniques, organization, and labor relations are occurring simultaneously. The course of the division and integration of social labor is subject to dramatic and often unanticipated changes as industrial restructuring proceeds. As there are always several feasible paths, the direction taken is to a large degree unpredictable. There is no need to claim that the organizational ensembles put into place are the most efficient possible, only that they satisfy the general requirements of accumulation by generating dynamic economies over time. In the process of territorial-organizational transition, short-term events are critical to the unfolding of ultimate outcomes, for some become cemented into an

[17] In this instance, the industry did not relocate out of Los Angeles, but the microgeography of the complex was critically transformed.

emerging dominant organizational regime because they are inserted into a system of external economies. Other short-term experiments somehow are not reinforced, and are effectively dropped from further consideration. In other words, the relationship of short-term events to long-term outcomes is as critical for organizational-territorial regimes as it is for the development of industrial technologies. In both cases, development under conditions of disequilibrium does not lead to optimal outcomes which can be predicted from some set of data at the starting point; history, rather, is actively produced through the interplay of circumstance and human response.

5.3.4 From Fordist to post-Fordist territorial organization

Restructuring across a wide range of propulsive sectors may be referred to as crossing "industrial divides" (Piore and Sabel, 1984). In this view, new systems of production are put into place with quite different production technologies, labor relations and territorial-organizational frameworks. "Fordist mass production" was one such system, and it had some well-known characteristics:[18]

- large capital-intensive factories organized around either continuous flow processes (e.g., petrochemicals and steel) or assembly-line processes (e.g., cars, electrical machinery);
- production and marketing posited on standardization of output, long production runs, and dedicated, non-adaptable capital equipment;
- large firms internalizing large parts of production systems, seeking to enhance internal economies of scale;
- the factory of the large corporation occupying the critical position between upstream suppliers and downstream fabricators, who often manufacture using batch production methods;
- multi-plant oligopolistic corporations operating within a relatively stable competitive framework;
- industry-wide unions with national contracts.

Fordist mass production was formed geographically in a set of great manufacturing regions – such as the US Midwest, northern France and the Seine Valley, the Ruhr, and the Industrial Triangle of northern Italy. These regions were organized initially by propulsive lead firms surrounded by agglomerations of input – and service – providers, with a concomitant massing of working class populations (Murray, 1983). Typically, the territorial complex consisted of a large industrial metropolis surrounded by a network of smaller industrial cities.

[18] Further discussion of industrial ensembles and regimes of accumulation follows in chapter 7.

Over time, these formations have broken down, and a new system of "post-Fordist" production organization, with its own particular geographical foundations, appears to be emerging. The new meta-organizational framework for production is characterized by what may be called flexible production methods: rapid changeability of product and process specifications and configurations. Flexible production systems have the following broad characteristics compared with Fordism:

- more general purpose equipment and machinery, especially machines based around variable labor processes and/or programable computerized equipment;
- smaller, more specialized workplaces and firms, and greater reliance on subcontracting;
- greater attention to demand variations, to which the quantity and mix of inputs and outputs can be rapidly adjusted by altering procedures or the mix of participants;
- collective social and institutional order in place of hierarchical control exercised by the mass production corporation;
- more temporary and part-time hiring and more relaxed internal rules for assigning workers and managers to variable tasks.

While by no means universal, flexible production methods appear to be on the way to becoming dominant in the economies of the United States and Western Europe (Scott, 1988b). If Japanese methods can be considered "flexible mass production," flexible production systems can probably be called hegemonic in the capitalist world economy; but the distinctiveness of Japanese industrial organization makes us hesitant to force such a claim (Florida and Kenney, 1989).

Flexible production is marked by a decisive geographical re-concentration of production, and by the resurgence of the industrial district. All flexible production industries are marked by organizational fragmentation in which dense, unstandardized, transactional relations between firms are particularly important. Firms concentrate geographically in order to reduce the costs and difficulties of carrying out these transactions and to maximize their access to the cultural and informational context of the production district itself. Accordingly, we now find major concentrations of producer services in New York, London, and other world cities, together with dependent clusters in parts of their suburban fringes. We find new high-technology industrial districts scattered across North America and Western Europe, and revitalized craft industries in the Third Italy, in Los Angeles and New York, in Denmark, and in southern Germany (Storper and Scott, 1988). Because there is ongoing secular development of the organizational capabilities of capitalist economies as a whole, these are neither territorially nor organizationally equivalent to the industrial districts of the nineteenth century (contrast Piore and Sabel, 1984): they are much less self-contained, being situated within wider and deeper regional, national, and international divisions of labor. Indeed, it is precisely their insertion into this richer global division of social labor that permits the respecialization of production upon which the new type of territorial organization is premised.

6 Labor – The Politics of Place and Workplace

6.0 The labor "factor" in location

One of the most glaring shortcomings of traditional economic theory is the treatment of labor as just another "factor of production." The same error persists in location theory, where labor is viewed as important only in a few labor-intensive industries (e.g., Fuchs, 1962). Our purpose here is to enrich this impoverished view, and to push the labor "factor" to the forefront in our analysis of the geography of industrial capitalism.

It is readily apparent that closure of unionized plants in the northern United States has long borne some relation to the construction of new factories in the non-union states of the Sunbelt (McLaughlin and Robock, 1949; Peet, 1984). The movement of electronic assembly plants to Asia cannot be understood apart from the almost one-to-ten wage ratio between there and the United States. Yet all manufacturing has not been relocated in the South or abroad simply because wages are so much lower; industrial shifts cannot be explained in terms of a labor factor measured only by wages and unionization rates. Nonetheless, labor matters fundamentally to the locational specifications of industry. Its role is misjudged in large part because the wrong aspects of labor have been measured – by both those who believe labor is a critical variable and those who do not.

To appreciate the role of labor in location, it is necessary to rethink quite thoroughly the conventional (neoclassical) notion of labor as a commodity possessing given qualities (skills), exchanged in labor markets for a wage, and utilized by industry in such a way as to optimize its marginal product. To this end, we must dissect the labor exchange with care, inquiring into such matters as its noncommodity character, its geographical availability, the distinctive technological demands of different industries, the functioning of labor markets, the participation of labor in the workplace, and the contradictory character of employment. We can then reconsider industrial location as a managerial strategy for dealing with labor, and the broader significance of the interaction of employment and location over time. What results is a reinterpretation of the meaning of the term "spatial division of labor" as a social process rather than a managerial allocation of industrial facilities to predestined locations.

6.1 The unique character of labor supplies

Laborforces exhibit a high degree of geographical differentiation owing to labor's unique nature. Labor differs fundamentally from real commodities because it is embodied in living, conscious human beings and because human activity (work) is an irreducible, ubiquitous feature of human existence and social life (Marx, 1867). Nonetheless, neoclassical economists and location theorists treat labor in the same terms as real commodity inputs and outputs, that is, as reducible to price (wages) and quality (skills). True commodities can be industrially produced, purchased at a consistent price and standard quality, owned outright by the purchaser (or rented) and employed in a strictly technical manner; their prices are subject to a geographical levelling process, and their markets can be spatially integrated over long distances. Labor, in contrast, remains idiosyncratic and place-bound because none of these conditions holds. Labor takes a commodity form, but it is not a true commodity.[1] Labor is a pseudo-commodity with four distinct dimensions: conditions of purchase; performance capacity; actual performance; and reproduction in place.

Conditions of purchase Labor must be purchased on the market at a price, the wage. Wages are the largest single element of input costs on the average for all industries, and a significant item for every single industry (Lynch, 1973; Moriarty, 1977; USBEA, 1978).[2] The wage bill need not be the largest factor in unit output costs to be the most important in choosing a location in any case, because cost differences are what count, and these depend on input price variations between places. Inter-regional wage differentials remain substantial in the United States and are increasing (Browne, 1987); and international wage differentials are generally much greater. The price of labor also includes the "social wage," such things as health benefits, retirement plans, and safety protections, which account for about one-third of the wage package on average in the United States and Western Europe. Most aspects of the social wage exhibit even greater inter-regional variation than nominal wages (Bluestone and Harrison, 1982).

[1] The commodity form is called "labor-power" by Marxists, but we retain the ordinary – if sometimes ambiguous – usage of the word labor.

[2] The share of wages in total input costs for manufacturing has declined, owing to mechanization, but the decline has not been as precipitous as the general level of technical advance might suggest because it has been offset by new tasks, rising real wages, and a decline in unit costs of materials and machinery; as a result, non-production workers have not substantially replaced production workers in basic manufacturing (Singlemann and Tienda, 1985). The labor-intensity of office, sales, or other non-manufacturing work is little in dispute.

Furthermore, labor is not a fully variable factor of production. Wage contracts are always implicitly obligational. The same job in two regions may carry very different expectations for the workers' inter-temporal income profile, for the security and regularity of employment, and so on. The degree of obligation exhibits great inter-occupational, inter-industrial, and inter-regional variation (Wachter and Williamson, 1978; Clark, 1985). Perhaps the most important aspect of the obligational nature of labor contracts is the attempt to regulate the variability of labor use through work rules, hiring and firing rules, and so on, and these, too, vary according to place-bound legislation (OECD, 1986). The delicate social nature of the relationship between workers and employers cannot be reduced to the simple contractual transactions that govern other inputs; nominal wages, social wages, and obligational relations are relatively slow to adapt to changes in labor demand (Clark et al., 1986).

Performance capacity Performance capacity includes not only technical skill, but other necessary capacities for effective labor, such as creativity, patience, self-direction, adaptability, emotional stability and more. The performance capacity of workforces is known to vary markedly between regions (Scoville, 1973). Particular laborforce skills are frequently associated with particular places: in Britain for instance, one finds metalworkers in the Midlands, office professionals in London, miners in South Wales, and academics in Oxford (Massey and Meegan, 1978). Workers in automobile factories can not immediately shift to assembling integrated circuits nor can Detroit's engineers move easily into advanced computer design.[3]

Performance and labor control The most fundamental difference between labor and true commodities is that there is no guarantee you get what you pay for (Gintis and Bowles, 1981). Performance capacity is not the same as actual performance because the worker can consciously limit or otherwise regulate work effort on the job. Workers, unlike machines, must be willing to engage their capacity for work; they have the power to resist being used by the capitalist. The intensity, continuity and quality of work that can be elicited from workers with a given degree of supervision, monitoring, and punishment, is of fundamental importance to the employer. The usual term for this problem is "labor control" (Edwards, 1979). Control is a two-edged sword, however, since it is not enough that workers simply follow the employer's orders; they must actively participate in production (Burawoy, 1979). Capitalist production contains a fundamental tension between the need for labor control and the need for creative participation. Labor control is a basic consideration in location

[3] As an industry's level of output and standardization increases, skill levels and distinctiveness may decrease, but not as much as is often thought (see section 6.2.2.).

decisions, because laborforces vary widely in their habits, expectations and norms about work and rewards (Durand and Durand, 1971; Low-Beer, 1978; Schmenner, 1978; Gallie, 1979).

Reproduction in place The dimensions of labor supply differentiation just discussed are socially produced, largely *in situ*. In part, they are residuals of local and regional histories that predate industrialization; but in advanced capitalist nations, industrial history is the principal source of differentiation. Because industries offer very different conditions of employment, workforces build up diverse historical standards of living, skills, norms, and work experience. This differentiation has a powerful impact on social reproduction in local communities, even though workers are free to leave the plant at day's end to pursue their own lives beyond the reach of the employer (Dawley, 1976; Hirsch, 1978; Walkowitz, 1978; Hareven, 1982).

Local labor markets deserve special emphasis because of labor's relative day-by-day immobility which gives an irreducible role to place-bound homes and communities. For the vast majority of workers, the place of employment lies within the range of the daily journey to work and back (Berry, 1972). It takes time and spatial propinquity for the central institutions of daily life - family, church, clubs, schools, sports teams, union locals, etc. - to take shape (Bott, 1971; Fischer, 1976). Once established, these outlive individual participants to benefit, and be sustained by, generations of workers. The result is a fabric of distinctive, lasting local communities and cultures woven into the landscape of labor, even in the highly mobile environment of the US working class (Timms, 1971; Warren, 1978). The form, purpose and location of capital can generally be transformed with greater speed than the supply, location or performance of whole working communities, or social collectivities, of labor. As such, communities form a landscape of varied possibility for capitalists seeking labor.

Long-distance immigrants, permanent migratory workers, and seasonal migrants can, of course, supply significant segments of the laboring population (Burawoy, 1976; Piore, 1979; DeJanvry, 1981). Nonetheless, migration seldom eliminates inter-regional differences in wages, skills, work attitudes, or cultural traditions. Because labor's institutions usually outlive individual workers, local conditions are transmitted to newcomers. Where mass migration does occur, the new arrivals tend not to blend in but to be sharply segmented, creating new labor markets in the midst of old, as in the immigrant communities of US cities (Sassen, 1988; Zunz, 1982; Morales, 1984).[4]

[4] Local conditions of labor reproduction can be changed, to be sure, but such transitions are time-consuming and costly to employers (see section 6.5.3). Employers of new immigrant labor, or employers who move to peripheral areas in search of new labor supplies, sometimes discover their workers are not as malleable as they had hoped, as Italian industrialists found in hiring southern workers for northern auto plants in the 1960s (Low-Beer, 1978; Gallie, 1979; Sabel, 1982).

6.2 Labor demand: payment, performance, and control

Having suggested why labor supply remains differentiated, we now look at labor demand, adding greater depth to the first three dimensions of labor: conditions of purchase, performance capacity, and actual performance.

6.2.1 Conditions of purchase: industry ability to pay

The universal forces of competition, class struggle, and accumulation of capital lead industries down divergent evolutionary paths, because each sector faces a fundamentally different set of possibilities and limits in its markets, technology, and organization, as argued previously. The nature of the product, the structure of markets, the possibilities for production technology, and production organization feed on each other to propel an industry along a particular development path; they engender divergent sectoral development, based on sequences of positive feedback between technical possibilities, cost structures, and market structures (Vietorisz and Harrison, 1973).

For example, airplane engines are a large, sophisticated product with limited markets, and this virtually insures the participation of large firms, governments, and international consortia. In automobiles, the massive costs of setting up assembly lines and a distribution network are major barriers to any firm seeking to enter the automobile industry, making this long the epitome of oligopoly (Sylos-Labini, 1962); at the same time, buoyant demand, mass production methods, and oligopolistic structure have insured high levels of return over long periods. Cotton textiles by contrast, are standardized and mass produced like cars, but demand has grown more slowly, and fashion shifts can be dramatic and frequent. Textiles cannot be automated much more at this time, and markets are large relative to the output of any one factory or firm. There is a continual opening for new competitors because machinery costs are low and distribution outlets widespread. Strong price competition coupled to unspectacular sales growth translates into modest profits.[5]

Divergent sectoral development gives rise to inter-industry wage variations by setting the upper limits to profit rates and wage levels. Different labor productivities are highly correlated to the value added by each worker, while favorable demand expansion and competitive conditions may allow for an additional measure of profits.[6] This

[5] These conditions can be dramatically altered by major technological breakthroughs or other systemic restructurings of an industry, of course.

[6] Extracted from "consumer surplus," in neoclassical parlance, from "inter-capitalist transfers of surplus value," in Marxist terms.

potential wage schedule represents the industry's ability-to-pay. It is hidden from view, however, because actual wages are a consequence of the division between labor and capital of the workers' value added and the industry's (or firm's) excess profits (Marx, 1867; Sraffa, 1960; Kalecki, 1971). That is, ability-to-pay becomes an actual wage schedule through class conflict over income shares.

6.2.2 Performance capacity: labor skills

Performance capacity affects both the productivity and quality of labor, and hence the ability-to-pay of any industry (or firm). It simultaneously affects the leverage of the worker, or groups of workers, to extract a better reward by virtue of contribution to and replaceability in the production process. Profits and wages are also influenced by the internal differentiation of labor demand within industries according to the varying performance capacities of workers across the division of labor. Performance capacity is usually treated under the more narrow rubric of "skill." Skill is a most difficult and complex phenomenon, which commands a vast and contradictory literature, both conceptual and empirical. We shall discuss only three questions to highlight the continued differentiation of labor skills. First, does modern industry tend to homogenize or differentiate skill demands as it mechanizes? Second, how does the division and combination of tasks develop and affect skill determination? Third, how is it that skills come to be socially valued?

Skill and mechanization Three basic positions have been articulated on how skill requirements tend to develop in the course of mechanization of capitalist industry: the radical labor process position, led by Braverman (1974); the post-industrialist skill upgrading school, epitomized by Bell (1973); and the new "post-Fordists," led by Piore and Sabel (1984). Our analysis of mechanization leads to conclusions at odds with the prevailing wisdom about the effect of machinery on labor.

The Marxist critique, given new life by Braverman, has been immensely influential in labor process studies. It holds that automation degrades labor, leaving deskilled or unemployed workers in its wake. In geographical terms, this idea translates in the production-maturation version of the product cycle: as an industry mechanizes, it reduces its dependence on skilled labor in central cities, allowing it to seek out cheaper, unskilled laborforces in less developed rural areas. Industry decentralizes as it mechanizes. Marx and Braverman identified a profound movement toward the transformation of work under the direction of capital, with the attendant destruction of craft work: tasks have been divided into minutiae, each motion rationalized, and work transferred to machines in industry after industry, job after job. Generations of workers have found their skills – and wages – cut out from under them (Montgomery, 1979; Noble,

1986). Braverman's view continues to be reconfirmed in a surprising variety of industries and places (Zimbalist, 1979); of special importance in the 1980s has been the Taylorization of office work and consumer service industries (Crompton and Jones, 1984; Baran, 1986). But this is only a partial view of mechanization, as a number of observers have recently pointed out (Wood, 1982, 1988). Labor, as the elemental human factor in production and the embodiment of the human capacity for knowledge and action, can never be entirely replaced; it reappears amongst and around the machines and reasserts itself.

There are several reasons for the resistance of the labor process to degradation (Walker, 1988b). First, many jobs are poorly mechanized, despite all progress in the mechanical arts. Machines are simply not as good at many tasks as human beings, given the the suppleness of the human hand, the perception of the human eye, the easy communication of the human voice and the fine judgement of the human mind. Materials are refractory, products tricky to mass produce. Even where mechanization has gone forward in some areas of production, there are innumerable labor-intensive interstices that resist rationalization (Nichols and Beynon, 1977). Second, while some improvements move the labor into the machine, others put the machine back in the hand of the worker. Automatic tools such as drill presses, jackhammers and personal computers enhance the capabilities of workers in a straightforward way. Third, mechanization is not unidimensional (Bell, 1972). The job structure may include different levels of skill according to which dimension is considered. There are also shifts and reverses in direction: capital may have tried to reduce labor to nothing in one dimension, e.g., mechanical transfer, only to find that this makes the work more difficult in another, e.g., work integration. Progress may demand the reintroduction of "lost" labor and skills, as Japanese production methods have shown (Sayer, 1986b). Fourth, mechanization is accompanied by the creation of new tasks, which are frequently ill-defined, creative, or otherwise require human attention. These include such auxiliary work as product design and machine repair. For example, programming a computer to run a numerically-controlled machine tool involves considerable skill. (The work may, of course, go to another worker – an engineer instead of a machinist – but this has to do with the allocation of work rather than its technical parameters.)

Counterposed to the deskilling hypothesis is a view as optimistic as Braverman's was pessimistic. This is the notion that automation frees workers for activities of a higher order, manifested in a shifting upward of the entire occupational structure (Kuznets, 1957; Blauner, 1964; Bell, 1973). Putting aside utopian notions, however, there is a real basis for seeing an upgrading of industrial work. For one thing, it is possible to build more productive systems by using and developing the minds of workers. As Sayer argues about Japanese production methods, "just-in-time is not simply a low-inventory system of production . . . It is a particular and sophisticated method of learning-by-doing" (1986b, p. 53).[7] This point is buttressed by the recent experiences of several major manufacturing firms. Fiat, Olivetti and General Motors all attempted to adopt extremely automated forms of production, yet each discovered that greater productivity advances could be had by the social reorganization of production toward a flexible, learning-by-doing model (Locke, 1987). The GM-Toyota joint-venture plant in Fremont, California is GM's most productive, yet its level of mechanical and electronic sophistication is lower than that of many other plants in the firm (*Business Week*, 1987).

Machines also create skills by making possible forms of work that never previously existed, involving new knowledge about materials, machines and products: airplane pilots, computer operators and audio technicians have no direct antecedents (Adler, 1985). Machine culture has generally put everyone on a higher plane of technical understanding. Furthermore, creativity and responsibility are at a premium in the handling of immense machine systems whose malfunctioning can be terribly costly. The "new and unexpected ways of failing" of automated systems means that machines increasingly need workers who can oversee their functioning and cope with breakdown in a creative way (Hirschhorn, 1984). Modern automation can involve a gradual upgrading of skill because it reverses the extreme task fragmentation and subordination to machine rhythms found in early mechanization.

The latter point has some overlap with the "post-Fordist" view of skill. Post-Fordist (or post-Taylorist) production, it is argued, requires constant change and innovation in products, and thus necessitates that workers possess general skills to work with flexibly specialized equipment and to adapt to a wide range of work situations. A new kind of skill is the ability to move between tasks relying on general conceptions of the technology and work involved. These "polyvalent skills" have grown as "craft skills," i.e., knowing many sides of one specific job, have declined (Coriat, 1979). Post-Fordist production increases skill demands with its broader job classifications and permissible work assignments, with forms of greater worker participation in production planning and shop management, by moving from compensation rates based on job assignment to personal rates based on knowledge level, and so on (Piore, 1987).

Trying to determine the true relationship between mechanization and skill is, perhaps, a futile quest. The perspectives of deskilling, upgrading, and generalization emanate, at heart, from looking at different groups of

[7] Sayer continues: "And this is a primary reason why Japanese firms have had so much success in out-competing established western firms which had treated their industries as mature and for which relocation to cheap labour countries was seen as the only way of improving competitiveness. Whereas the 'learning curves' of western firms in consumer electronics, air conditioners, cars and office machinery had flattened out, those of the Japanese continued to improve."

industries: Fordist mass production, advanced engineering, and craft-like batch production, respectively. It is difficult to generalize from any one of these to the innate logic of capitalism or industrialism. Skill demands have been, and will continue to be, differentiated because of the diversity, product-specific nature, and dynamic histories of different sectors. With the growth of new sectors, "outdated" moments may be rediscovered, as in the Taylorization of office work or the upgrading of factory jobs through computer-aided machinery.

Skill and the division of labor The classic critiques of work under capitalism refer also to the division of labor (Smith 1776; Marx, 1867). In differentiating particular tasks, the division of labor often creates skills through the mastering of new specializations (the flute player in an orchestra is, after all, a kind of detail worker); carried to extremes (in the manner of Taylor), however, the detail division of labor is a way of reducing tasks to the bare minimum of human action and creative engagement.

Labor division and job specialization are controlled not only by machine technology, but by the scale and stability of markets. The extent of the market affects the mass of goods produced and hence the possibility for increased specialization, within the bounds of available technologies (Berger and Piore, 1981); the stability and uniformity of the market determine the degree of product standardization, and thus the ability to rivet the worker to repetitive tasks; these market characteristics depend, in turn, on the nature of the product – e.g., its uniqueness, durability and range of applications. For example, shoes change more rapidly in style and are less durable commodities than tractors; hence, the former are produced via batch production methods while the latter allow long production runs and a deepening of the division of labor. In the course of sectoral development, moreover, there is no unidirectional progression from less to more standardized, or stable demand (Storper and Christopherson, 1987). In short, substantial sectoral specificities will exist in every industry's division of social labor.

While product markets and mechanization define certain possibilities for dividing up a production process, and tasks have particular skill requirements, work tasks must be combined into jobs. It is jobs for which skills are demanded in labor markets. Tasks may be allocated to workers in more or less confined or expansive, equitable or uneven, ways. The scope of the job depends both on the internal organization of work within the workplace and the external organization of production units into large or small firms, industries and territorial clusters. Job specialization or generalization, variability or repetitiveness, depend on what role the worker, workplace, or firm is allocated in a larger production system.

Some view the division of labor, and hence skill, as structured by social relations while others see technological determination at work. Neither view is correct. Not only is technical change itself partly structured by class relations and capitalist choices, but in the end machinery does not narrowly determine how production is divided and integrated. Conversely, it would be unwise to see class relations as directly expressing themselves in the division of labor; that division stands on its own as a fundamental dimension of the forces of production and hence of skill (Walker, 1985b; Cohen, 1987).

Social valuation and skill The skills debate is due, in part, to the difficulty of agreement on terms. Braverman and his followers hold that worker autonomy is critical, but there is no unambiguous definition of the autonomous worker (Wright, 1985): is a worker who has great discretion over a small batch operation more skilled than the polyvalent laborer in continuous process manufacturing? Others, such as Spenner (1983), have advanced "substantive complexity" as the measure of skill, as indicated by the number and difficulty of tasks, the expected standards of proficiency, and so on; yet this definition misses the social dimensions of work, especially the skills required for integration of the collective laborer. Finally, all such theories are plagued by the fact that skills are incommensurable over time: the most important changes in skill are often qualitative, as new products and processes introduce tasks that cannot easily be arrayed on an ordinal scale (Adler, 1985).

We may not, in the end, be able to escape from a kind of hermeneutic circle with respect to the evaluation of skill (cf. Sayer, 1984). The neoclassical position on skill and wages can help us clarify this idea. Neoclassicals deal with skill through a straightforward market-clearing model: wages are the equilibrium rentals of skilled labor. Alternatively, they argue that investment in "human capital" augments supplies of scarce labor, thus adding an inter-temporal perspective on skill, and creating a production function for it (Becker, 1964). Either way, efficiency wages result and there is a monotonic relationship between skills and their rewards, because higher skills mean higher productivity at the margin. The neoclassicals score a certain point by observing that wage differentials are, in part, society's payments to scarcity of performance capacities and willingness to work hard for an employer; they are a quasi-rent which workers extract by virtue of their command of a scarce resource.

This suggests that skill is a social construct, and no single definition could ever capture it. As such, skill results from the interplay of quantitative (supply-demand) and qualitative (ideological, cultural, political) forces. Skill is that which is problematic, or difficult to secure, or – and this is equally important – perceived as difficult to secure. We need not be troubled by the untrue parts of the neoclassical perspective: since market clearing and substitution are not the centerpiece of economic history, skill valuations are not objectively determined by market forces. In other words, skill is socially constructed through processes of social production, which are coupled to ongoing processes of social evaluation. As a result, very few wage levels hit the mark in any objective sense.

The social production of skill comes through all manner of circumstance, from formal certification to on-the-job training, and is a dynamic process. For instance, what is considered a skill in one period may become the norm in another (reading, writing, driving a car), and the norm commands little rental above subsistence; for this reason, general technological upgrading of work may have little positive effect on skill valuation: a truck driver is still a "teamster." Many jobs are valued according to who holds those jobs rather than the nature of the work itself: for example, garment assembly is, in general, more complex than car assembly, but the women who work in garments are considered less skilled than men who work in car factories. The subordinate position of women means that whatever they do is labelled as unskilled, while patriarchy allows over-valuing of men's skills (Phillips and Taylor, 1980; Walby, 1986). The restrictive practices and ideology of professional and technical workers have the same effect, as in the limited access to, and over-valuation of, medical and law degrees in US society (Starr, 1982). Indeed, one could go so far as to say that, overall, the degradation of work in capitalist societies is a result of the social devaluation of the working class in general, not the other way around – as was true of previous class societies.

Skills are those qualities that employers need but cannot perfectly control because they cannot produce them in any reliable or economical way. Workers gain bargaining power from possession of these qualities. This is decidedly not a return to human capital theory, since education, training and worker formation in general do not function via market clearing, and because institutions and politics influence workers' bargaining power in labor markets. This is why statistical studies of occupational and industry wage structures always show a large residual (often greater than 50 percent) after accounting for labor productivity, industry profitability, and skill measures such as experience and training (OECD, 1986). Something is missing. In what follows, we suggest that skill is merely an entry point to a more complex set of social relations in the workplace and the labor market that ultimately determine the conditions of employment.

6.2.3 Performance control: relations in production

Production rests on an objective basis in the sense that certain operations must be performed on materials to produce a particular product. But the strict objectivity of production ends at the engineer's design table. Machines never run entirely by themselves; they require the intervention of workers. Because all workers retain some measure of control over the conception and execution of work, their active cooperation in the labor process is required (Wood, 1982). Managers must come to terms with workers if anything is to be accomplished and this sometimes gives workers leverage in bargaining, above and beyond their mere scarcity value in the labor market. At the same time, workers need employment to survive and to have the opportunity to exercise their creative powers. Employers and workers are, in this sense, each other's captives in the workplace; their relationship is characterized by mutual dependency.

Moreover, production is always a collective process, involving communication, physical interaction, and group effort. Materials may have to be moved to and from work stations, as in machining; workers may be linked in a sequential work process, as in automobile assembly; groups of workers may come together to assemble a particular large item, as in aircraft assembly; machine operators may need to communicate and coordinate functions, as in petrochemicals; or workers may simply be confined in a small space, as in garment factories (Storper, 1982). In short, some social interaction is based on patterns of technical interdependence specific to each industry. The quality of the interaction also varies, depending on the industry's division of labor, which determines such things as work pace, noise, the need for mutual aid, and conjunction of work goals (Friedman, 1977). Even non-worktime socializing in lunchrooms and bars is affected by shift schedules, common experiences at work, and exhaustion.

The combined result of the individual and collective activity of work is not only the production of commodities, but the creation of a social life, which may provide collective strength, and of a social consciousness, which may create a sense of worth and resistance to exploitation.[8] Industrial sociologists have frequently pointed out that the organization of work – the degree of job autonomy, worker interaction, group segmentation, and so on – has a powerful influence on whether workers come to understand the employer's dependence on them and translate this into demands for higher rewards (Burawoy, 1979). An industry's division of labor also determines the material bases for workers' control over the production process (Zeitlin, 1979). Worker militancy is thus the result of two basic influences: the condition of the worker in the political economy (in terms of wages, historical traditions, and freedom to organize) and the character of the production process itself (Woodward, 1969). Neither of these alone allows us to predict worker or employer action, but together they do establish parameters for it (Sabel, 1982; Burawoy, 1985).

For example, textiles and auto assembly both involve mechanized, routine, semiskilled labor, yet auto workers are often more unionized, more militant and better paid. This reflects not only their employers' greater ability-to-pay, but also a distinctive organization of work that helps generate solidarity: the common status of the workers, the common pace of the assembly line (which it is in every worker's interest to slow down), stationary positions which allow conversation with others nearby, close coordination of several tasks, and the common practice of helping out the next worker in the line (Beynon, 1973). In textile factories, on the

[8] By the same token, labor history is replete with examples of solidarity of one group of workers in opposition to others along lines of occupation, race, gender, and ethnicity.

other hand, workers are divided by the extreme din, faster pace, shifting work position (moving among machines), and a competition that results from the fact that interrelated tasks can be performed at different paces. As a result, the workforce is sharply segmented along task distinctions: weavers, machine fixers, doffers, and slashers (and the wholly separate work groups in carding and dyeing) (Blauner, 1964).

Cognizant of the potential of workers to develop solidarity or to resist (passively or actively), employers use counteracting strategies and construct systems of social control within the workplace. These include enhancing the power of the foreman, offering individual rewards, organizing work at the micro-task level, internal labor market segmentation, and specific antisocializing rules (Edwards, 1979). There are important non-coercive methods of labor control as well, such as defining workers in terms of other social roles, as in the now-common practice of holding beauty contests and cosmetic sales in the workplace in Asian electronics factories (Ong, 1987). Performance control is thus a key element in the relations between workers and employers. Workers and managers together manufacture social worlds in miniature in the workplace, the firm and across industries. No one creates these worlds entirely by design.[9] The social relations of the workplace, or what Burawoy (1981) calls "relations-in-production," have qualities peculiar to every industry or group of industries, whatever the levelling effect of labor market legislation and large unions.

6.3 The employment relation

We have thus far shown that labor demand has three fundamental determinants, each with industry-specific qualities: ability-to-pay, technical features of the labor process, and relations in production. But this model is too static: sectoral conditions do not set up a determinate labor demand so much as they set the bounds for a political relationship between employers and workers which serves as an ongoing basis for employers' definitions of labor demand and for a distributional battle between workers and capitalists. The politics of production (Burawoy, 1985) need to be set within the larger frame of labor markets and broader processes of social reproduction.[10] People in different labor market or national contexts will practice production politics differently. Employers and workers use the conditions in the labor market and in the larger community and society as a way of influencing life within the workplace. This wider relationship between workers and employers is the employment relation, and it is only in the context of employment relations that real demands for labor are defined and supplies of labor produced.

[9] Indeed, capitalists and laborers become so ensnared in these relations, or social games, that the class nature of production and sometimes even the productive goal itself may be lost from sight (Burawoy, 1979).

[10] In the narrow terms of wage determination theory, even if one were to account for all the dimensions of labor demand elucidated thus far - productivity, profitability, skill, and control - considerable unexplained residuals would remain.

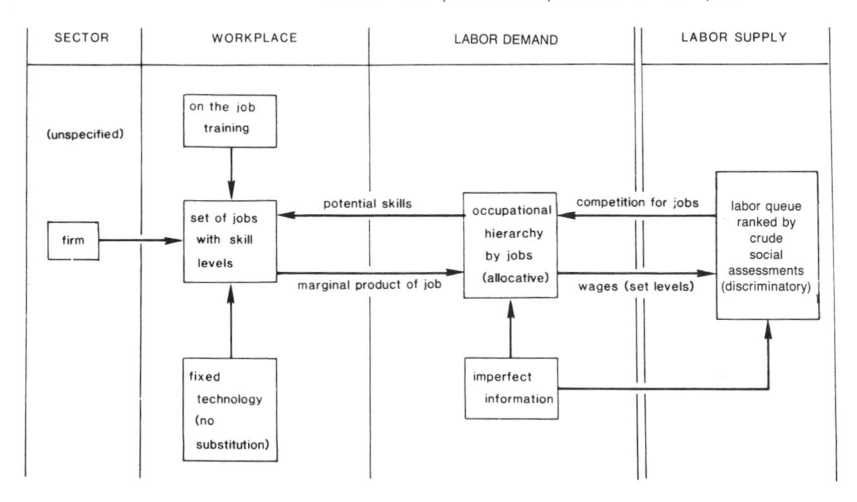

Figure 6.1 *Labor queue model* In this model, employers operating under imperfect information about potential employees and their skills use crude social indicators as a shorthand for training costs, worker performance, and managerial control problems.

6.3.1 Labor queues

Production in capitalist industry rests on reconciling employers' contradictory requirements for performance and control from the same group of workers. Management attempts to keep rewards within the bounds of profitability while wresting productivity from workers. At the same time, employers must maintain managerial prerogatives so they can respond to changes introduced by competitors. In seeking to maintain this essentially undemocratic and exploitive condition, management must not encourage too much resistance or resentment (Bowles and Gintis, 1986). One way employers do this is to select labor carefully from the external labor market. Previous employment and non-work experience therefore become important to the political propensities that workers bring to the workplace (Sabel, 1982; Massey, 1984). Different workers make for different outcomes, all other things being equal.

In conventional terms, this amounts to a modified "labor queue" model of labor demand (Thurow, 1975) (see figure 6.1). In this model, jobs come with known characteristics and employers pick and train workers so they can generate normal levels of productivity and profits. Employers rank labor supplies on the basis of potential control versus performance costs, thereby establishing a labor queue. Lacking information on individual workers, they must use indirect indicators, and therefore employers try to identify background characteristics that are good predictors: educational level, for example, may indicate an ability to absorb training, and success in school requires a kind of work discipline. For most industrial jobs, however, individuals are classified according to features descriptive of the larger population, such as race, gender, age, appearance, psychological attributes, or IQ. This forms the basis of systematic discrimination in the labor market.

6.3.2 The formation of employment relations

The labor queue model captures the arbitrary element of job allocation and employer power, but provides little sense of how concessions are ever made to labor. With careful discrimination, employers should be able to resolve the tensions between control and performance simply by inserting the right social raw material into the workplace – but they do not always succeed. The dynamics of worker-employer politics require consideration in terms of a model of bilateral employment relations.

No matter how skillfully employers define their ideal labor queues, they must accommodate to prevailing labor market conditions, worker struggles, and competitive shifts. For example, even where capitalists target exploitable populations, a successful organizing drive or strike may blunt the employer's initial advantage. In sectors where employers are forced to concede substantial rewards to workers, however, they are likely to select groups who are already more privileged, and therefore seem to "fit" their enhanced position better (Rubery, 1980; Walby, 1986); often with the active collaboration of those groups, especially white males (Hartman, 1979). More often than not, past advantage translates into greater solidarity and leverage that help maintain a favored group's position in the labor market. In any case, there is a subtle, historical interaction of demand and supply. Any real labor queue is therefore premised on an employer's prior knowledge of what can and cannot be done in a labor market, given its political and institutional history. The first semiconductor plants to go to Southeast Asia used the experience of textile firms which had preceded them, as well as their own knowledge of job allocation in the United States (Ong, 1987).

A predetermined labor demand exists only at the moment a wholly new laborforce is hired and production begins afresh. That moment is extinguished as soon as the living processes of labor exchange and production are underway. The labor exchange must in effect be renewed each day. In production, labor and capital come together in a way that is at once a market transaction, a labor process, and a scene of daily life: that is the employment relation. Furthermore, because employers and workers are captives of each other and neither side is free to get all it wants from the employment bargain, production is successfully undertaken within an unfolding process of negotiation. If, for example, the firm's competitive status depends on product quality or reliable delivery, as in the computer industry, certain concessions are required (Kalleberg and Griffin, 1978; Freeman, 1978). These affect both rewards in the labor market, and work rules and practices inside the factory. If the employer has little or no latitude, and if its performance demands are relatively low, as with cotton textiles, then employers must attempt to curb workers' demands (Storper, 1982). Just how much employers must yield varies according to the nature of the production process and the configurations of social power and solidarity in the workplace and labor market. Employment relations are ultimately the indeterminate result of human agency. The result is a dynamic process of definition and redefinition of labor queues due to the practice of employment relations (see figure 6.2).[11]

6.3.3 Employment relations and labor market segmentation

Differentiated employment relations produce a series of occupation- and industry-specific labor queues that define hierarchical labor market segments between which there are strong institutional and traditional barriers and thus relatively little mobility (Edwards et al., 1975; Wilkinson, 1981). As a result, labor rewards are not evenly or proportionally related to performance across industries and occupations. Some workers are able to secure privileged conditions, while others are paid less than warranted by the performance demanded of them. In neoclassical economics, fair exchange is the rule, whether for doorknobs or for labor. Thus, labor economics, particularly its sophisticated human capital variant, argues that workers are paid the value of their marginal product, whether measured by skills or effort (Becker, 1964). Empirical studies of labor market segmentation have not borne out these theories, however. Time after time, rewards – including wages, work condition, stability, advancement, and autonomy – have been found not to coincide with any of the major measures of workers' marginal productivity (Reich, 1981).[12] There are systematic industry- and occupation-specific differences in rewards even after one corrects for skill, training time, cost, effort, and availability. In the United States, the spread of wages and working conditions has long been much wider than the spread of performance capabilities (Scoville, 1973). Labor market segmentation provides a clue to solving the riddle of wage structures.[13] When we bring together an enriched socio-technical

[11] Job queues may also have a long-term feedback effect on the division of labor: the possibility of dividing the labor force into groups defined by race, gender, age, and nationality may afford employers opportunities to arrange the division of labor in production in ways that would otherwise not be possible.

[12] Even neoclassical economists sometimes concede this fact (Macdonald and Solow, 1985).

[13] Various attempts have been made to accommodate labor market segmentation to neoclassical theories of wage determination. One view attributes such segmentation to the costs of transactions, where labor queues minimize information and transaction costs (Arrow, 1972). Others claim that labor market segmentation is the frictional outcome of asymmetrical wage adjustments and will be overcome in time. Through labor absorption and release mechanisms, wages ultimately must be bargained back to a level at which asymmetries (segmentation) are eliminated. A more flexible neoclassical response holds that labor market segmentation redistributes wages between different groups of workers just as market imperfections redistribute profits among capitalists (quasi-rents). Indeed, the two are said to be linked, as, for example, in the strong correspondence of capital-intensive industries and relatively high wages. This is true because physical capital has a quasi-fixed relationship to high skill requirements (Griliches, 1969) and because skills are capitalized in the form of investments in worker training. Capital-skill complementarity, then, can lead to segmented

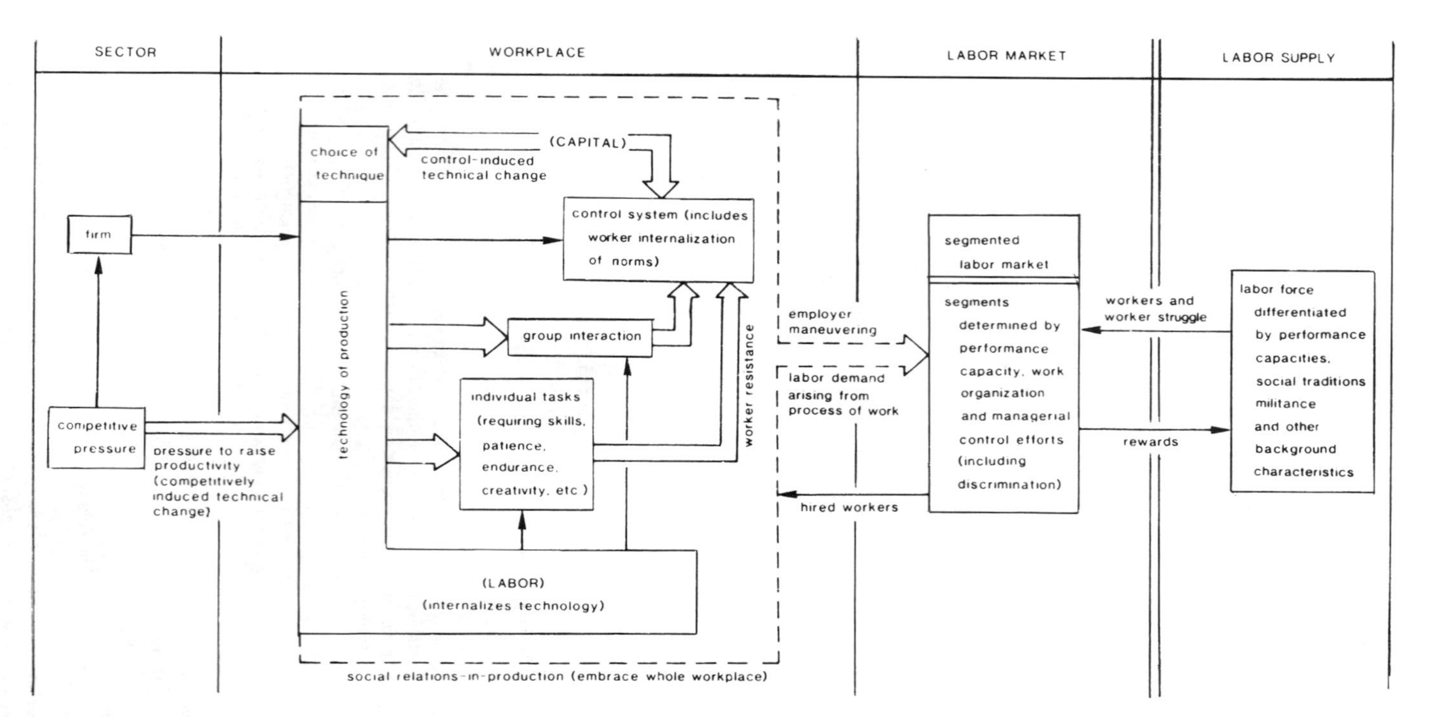

Figure 6.2 *Bilateral employment model* In this model, managers and workers are involved in ongoing conflict and conciliation over the terms of work. Employers seek to control labor and to keep down wages, and their maneuvering leads to segmentation of labor markets to divide and weaken the working class. Workers, on their part, come to the job with prior experience and traditions of militance, and use their toeholds of power in the workplace, drawn from individual skills and group interaction, to extract better working conditions and wages from management; this, too, may contribute to labor market segmentation. The social relations in the workplace have a certain life of their own, while competitively-induced technical change may systematically alter the shape of the labor process over time

conception of labor demand with real labor supplies through the practice of employment relations, segmented wage patterns result. This means that there is no possibility of a fully-determinate *ex ante* model of wage formation; politics is not a fully-determinate business, even when the participants' circumstances and backgrounds are known. In our view, segmentation, and its effects on income distribution, is an unintended consequence of the conflict between labor and capital within socially- and technically-defined strategic conditions (Gintis and Bowles, 1981), rather than any functionally-determined or lawful outcome of markets or of class relations as a whole.

Labor market segmentation in the post-war economies of the United States and Western Europe until the mid-1970s had the following broad configuration:

- An "independent primary" segment featured jobs with high wages and full-time, full-year employment, with work usually self-directed and, to some extent, self-controlled. The worker was relatively autonomous and enjoyed a considerable amount of social status. Professionals and skilled craft workers fell into this category, and tended to be white males.
- A "subordinate primary" segment sported relatively high wages and full-time, full-year employment. Work was not self-directed, however, and workers faced occasional layoffs and limited chances for mobility. Advancement was attained through seniority-based wage increases and promotions in internal labor markets. Few brought marketable skills to the job; rather they gained skills through on-the-job training and experience.
- The "secondary" labor market contained jobs with low to moderate wages, where the worker typically faced the unpleasant prospect of limited mobility, overt social control on the job, instability, and physical discomfort.

In the postwar period, a large number of workers could be included in the subordinate primary segment. This has been attributed to the dominance of large batch or mass production industries with relatively high ability-to-pay, relatively favorable demand growth, and oligopolistic competition. These production systems generate a need for large numbers of "semi-skilled" workers who repeat narrowly-defined tasks. At the same time, these so-called "mass collective" workers were able to secure

---

wage patterns since some industries earn quasi-rents on capital which may be passed on in part to skilled workers (Sattlinger, 1975). These neoclassical efforts ignore technological and organizational change, as well as growth dynamics; they make no allowances for employer discrimination and political behavior on the part of capitalists (Dickens and Lang, 1985); they ignore the entire supply side of the labor market, including the essential nature of the labor factor and the dynamics of the employment relation (cf. Akerlof, 1982).

important concessions from employers through union and political action (Coriat, 1979; Murray, 1983; Davis, 1986). Labor market institutions, including internal labor markets, were constructed to regulate entry into favored positions, while labor contracts maintained peaceful industrial relations by converting stability into personal advancement through seniority, limiting the unions' shop-floor autonomy, and fixing conditions for several years at a time (Stone, 1981). In exchange, mass production firms were allowed to turn to small subcontractors for certain inputs at reduced scales of production, or to absorb temporary demand fluctuations, giving rise to a small firm sector which offered much lower wages and poorer working conditions (Berger and Piore, 1981). These secondary labor market jobs were (and continue to be) occupied principally by women, the young, or racial and ethnic minorities (Edwards et al., 1975).

6.3.4 Institutions and employment relations[14]

The employment relations that create labor market segmentation are never formed in a world of isolated, spontaneous relations in production and worker-employer politics; they depend on institutionalized practices (Blum, 1925; Dunlop, 1944, 1948; Clark, 1985, 1986). Systems designed to match labor supply and demand, and the resulting forms of labor market segmentation, are products of institutionally-guided action. The labor exchange is regulated by laws on minimum wages and benefits, open and closed shops and unemployment insurance, among others. Some laws are oriented to the labor process itself, such as health and safety regulations. Labor exchanges are shaped particularly by systems of contract law and practice. For example, US contracts tend to limit the unions' shop-floor autonomy while in Britain they confer discretion on the shop steward (Bok, 1971; Beynon, 1973). Contracts also influence labor practices by specifying the quantity of labor hired and the pattern of labor hoarding or firing over the course of time (OECD, 1986). Such contracts are products not only of bargaining positions, but of existing institutional frameworks and previous contractual arrangements (Klein, 1984). Indeed, current maneuvering over labor market "flexibility" and "rigidity" is being carried out precisely over the terms of these institutional rules.

Workers also search for jobs through institutionalized channels, from the extended family to the union hall, with all their historical peculiarities (Stigler, 1961, 1962; Harevan, 1982). These institutions powerfully influence the ways workers adjust to changes in labor demand (Clark et al., 1986). Recruitment practices are often sectorally specific, as when a government agency aids in supplying agricultural labor or the longshore union regulates the shape-up through a hiring hall. More broadly,

[14] This section was inspired by Moulaert (1987).

educational systems can be seen as institutionalized means of producing different kinds of workers (Bowles and Gintis, 1976).

Labor market institutions are geographically differentiated. Laws on unionization, worker protections, wage rates, benefits, training and the like are quite varied (Harrison, 1984; Morgan, 1985; Clark and Johnston, 1987). Moreover, even when the structure of institutions is theoretically integrated, as with US labor law, practices of interpretation and implementation vary a great deal from place to place (Clark, 1985; Johnston, 1986). Some of these differences attach to nation-states, or their subordinate jurisdictions, but many are the result of less formal practices, such as employer or working-class traditions. For example, some communities make more egalitarian wage demands than others (Clark, 1986). The degree of obligation in labor markets, and the institutional mechanisms by which obligations are enforced, thus exhibit great inter-regional variation (Clark, 1983). These differences may be reduced over time by the intervention of the state, national unions, or employers' organizations, but they may also grow more complexly differentiated over time as a reflection of local industrial history (Massey, 1984).

Nations thus have distinctive "factory regimes" that affect the outcome of production politics (Burawoy, 1985). These regimes differ not only in terms of substantive provisions of labor laws, but in degree of centralization of unions and employers, geographic uniformity, and position of the governing body in the state bureaucracy. In the United States, for example, a highly decentralized system of industrial relations leads to particularistic relations between specific employers, and workers groups, and thus divides unions from each other (Rogers, 1985; Clark, 1986).

Neoclassical approaches cannot account for institutional history, or the way that institutions interact with the strategies of various agents to produce or alter the institutional fabric (Hodgson, 1988). Institutions formalize political relations, and channel the power through established rules of discourse and practice (Giddens, 1984; Dworkin, 1985); these rules stabilize the outcome of power relations at particular moments in time and in particular places, and can therefore inhibit change (Rogers, 1985). Effective strategic action, then, requires maneuvering substantive agendas through institutional channels.

6.3.5 The dynamics of employment

As industries follow growth paths in which there are periodic restructurings, they inevitably disrupt employment relations. A change in the methods or organization of production can rupture existing arrangements in labor markets by altering the profit position of employers, the type of performance required, or the bargaining power of workers. If capitalists alter the division of labor, for example, it can disrupt alliances among groups of workers, open up new positions for women or minorities,

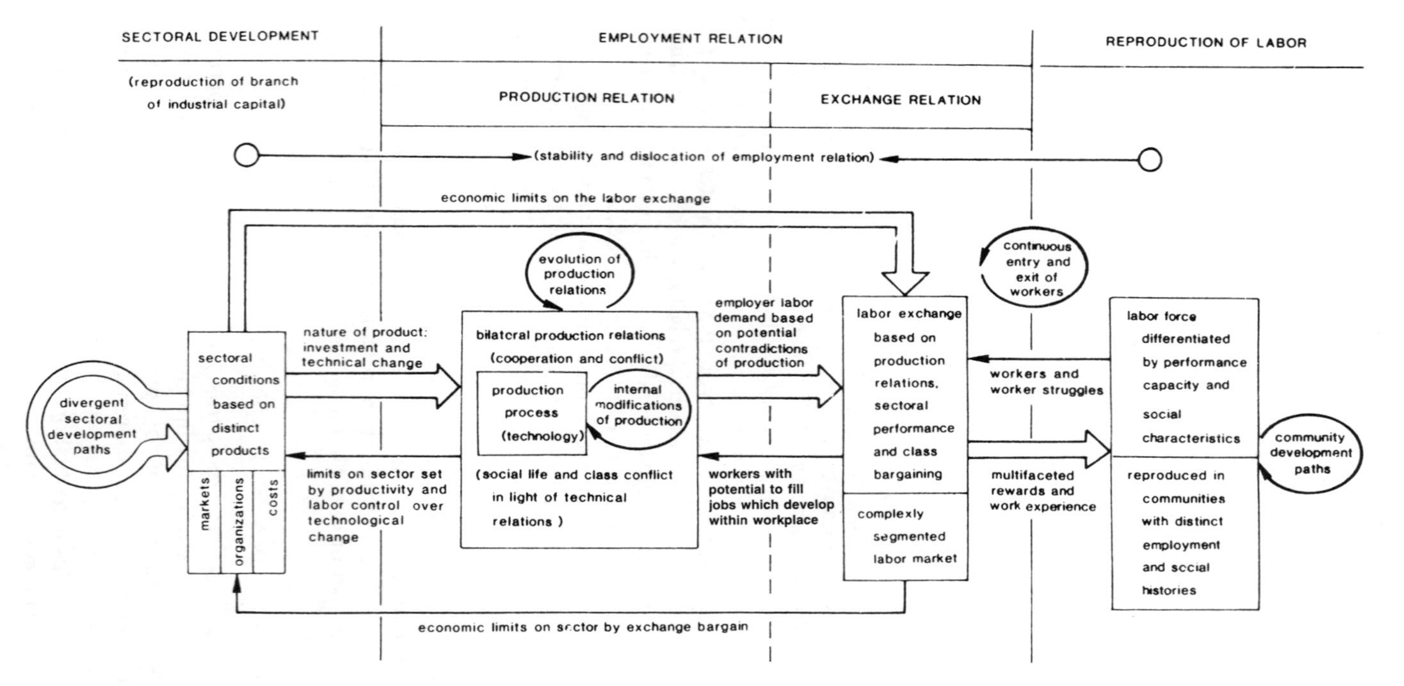

Figure 6.3 *Dynamic employment model* This model sets the bilateral employment model in motion, by putting the employment relation at the center of an ongoing interplay of industry development and social reproduction of labor. Sectors follow divergent and uneven growth paths, on the one hand, while communities are created and develop in relation to industries, but not as a simple reflection of the latter. The employment relation is a unity of production and exchange that generates complexly segmented labor markets and a spatial division of labor; and it is potentially unstable owing to the contradictory forces of accumulation and class impinging on it.

or redefine jobs, undermining workers' power in one labor market segment without improvement anywhere else. Conversely, alterations in labor markets, community life, or larger social and political movements can introduce disruptive activity from the workers' side, such as greater absenteeism or militant union organizing. A dynamic version of employment relations is depicted in figure 6.3.

A major question today is how contemporary methods of reorganizing industrial production are altering the shape of labor demand, employer practices, worker alliances, and labor markets (Ebel, 1985; Kelly, 1985). Employment levels in the Fordist mass production industries are clearly declining, and new types of subcontracting and outwork, with employment conditions typical of the secondary labor market segment, are increasing (Wilkinson, 1981; Rubery and Wilkinson, 1981; Solinas, 1982). These developments have been complemented by increasing numbers of secondary jobs in retailing and office-based industries (Freedman, 1976; Christopherson, 1986). At the same time, new kinds of semi-artisanal jobs have been generated in a variety of industries - including machine tools, textiles, and motion pictures - which sport relatively high wages even though workers face considerable insecurity (Christopherson and Storper, 1988). New contours of power in employment relations are being generated on a massive scale in the developed industrial economies. The institutions, organizations, and political alliances that burgeoned in older industrial areas under the Fordist regime of mass production are no longer operative over wide sectors of American and Western European societies. In new growth centers and regions of the United States, Europe, and Japan, novel social and political experiments in the deployment of productive labor are well under way. Undoubtedly, these new configurations of class power and technology will congeal into new contours of labor market segmentation (Harrison and Bluestone, 1988).

6.4 The geographical dynamics of employment relations

We must now transform our model of employment relations and labor market segmentation into a geographical context. Rethinking the employment relation means recasting the role of labor in industrial location. The problem is not simply to match plant labor demands to appropriate laborforces scattered over the landscape, as in Weberian theory (e.g., Weber, 1909; Hoover, 1948; Fuchs, 1968). Nor is location simply a strategy for management to manipulate laborforces and evade unions (Peet, 1984). Rather, employment relations are entwined in the geographical patterns of industrialization, previously enumerated. To understand this interaction, we must construct a picture of the way that local labor markets are built up, segmented and altered through the process of location and relocation of production.

6.4.1 Growth, localization, and employment relations

In the process of initial industry growth, a local growth complex comes into being, consisting of a variety of firms and factories linked by complex transactional relationships. As localized complexes of industrial production develop – as in San Jose, Lancashire, Seattle, or Dallas – they draw into their orbit a laborforce that embodies the skills and other attributes required by local employers. In the case of the new high-technology growth centers in the United States, the laborforce typically consists of two major segments: a stratum of highly-trained scientific and technical labor performing skilled work in research and development and advanced manufacturing; and a stratum of low-wage labor, usually comprised to a high degree of immigrants, who work in the many unskilled jobs. The micro-geography of Silicon Valley, for example, bears the strong imprint of this segmentation in patterns of residence and social reproduction (Saxenian, 1984).

In such place-bound communities, the process of learning the peculiar rhythms and imperatives of work are facilitated by the formation of a local working class culture, embodied in specific organizations such as churches or social clubs that reinforce legitimate expectations and habits. The immigrant community of Silicon Valley, for example, provides a rich cultural substrate of mutual suspicion, group support, sexism and class divisions from which electronic companies can draw to maintain labor segmentation, disunity and low wages (Cho, 1987; Hossfeld, 1988). Socialization is further enhanced where, as is typically the case, specialized educational institutions and training establishments are set up in the area to offer teaching programs in appropriate skills. Where these facilities are publicly provided or subsidized, producers avoid part of the costs of job training. In these and other ways, local labor markets function as intricate mechanisms that help sustain dynamic economies of industrial growth.

6.4.2 Locales and the politicization of employment relations

Community attitudes and practices can also be a medium of political response to the predicaments of work and life. The local population accumulates historical experience of its social condition as laborforce and citizenry, and its collective geographical presence eases the tasks of political mobilization around these roles. In the employment relation, workers confront management directly, and often the result is unionization, rising wages, more rigid work rules, and other abridgements of management's power to control production processes. But overt militancy and organization are not the workers' only source of strength. Contractual arrangements, such as seniority rules, may be sufficient to restrain employers' freedom. Workers also participate in specific traditions of occupational reproduction and differentiation that powerfully shape their performance capabilities at work and help shape the division of labor (Clark, 1986). In the arena of local politics, resistance depends on the wider balance of social forces in the community. Producers may be obliged to face rising production costs as a result of increased taxes on land, readjustments of zoning regulations, or tighter pollution controls. These community responses to growth constitute elements of an industrial politics of place (Foster, 1974; Dawley, 1976; Calhoun, 1982).

The politicization of workers partakes of "interdependence effects" in territorial labor markets (Hamermesh, 1975). That is, workers look to one another for guidance in how to respond to capital and how to define acceptable work conditions or living standards (Clark et al., 1986). Such wage interdependencies are the spatial equivalent of "pattern effects" among labor market segments (Dunlop, 1948). Spatial pattern effects encompass more than wages, however; they include the political element of employment relations inside and outside the workplace. Spatial pattern effects are inimical to labor market segmentation, so capitalists try to restrict the political unity of workers through a geographic strategy of divide and conquer (cf. Reich, 1981). They use the location and relocation of facilities as a maneuver against worker strength. As Clark, Gertler, and Whiteman point out, "the geographical division of labor into spatially discrete labor markets" is designed to "organize and exploit discrete sets of localized expectations;" it is "an active but surreptitious way of organizing the labor bargain or contract" (1986, p. 34). For example, scattering meatpacking plants in various small towns in the Plains States and Midwest makes it easier to play the union locals off against each other, as was demonstrated in the recent isolation and defeat of meatpackers in Austin, Minnesota, by Hormel Meat Company. Just to drive the point home, Hormel quickly closed a plant in Ottumwa, Iowa that had the temerity to support the Austin local.[15]

6.4.3 The politics of geographical shifts: growth centers

The formation and destruction of local employment relations provide the background for locational shifts. New and renewed industries not only give rise to their own specialized labor markets, they tend to relocate away from old centers, and even away from developed industrial areas altogether. We have discussed why such shifts are possible and even desirable in terms of industrial growth, technology, and organization; we now consider the ways in which the politics of place and workplace are part of the logic of industries choosing to move as technological windows of locational opportunity appear.

[15] Thanks to Brian Page for this example.

The politics of place pose significant problems for new industries as they emerge. In early phases, they are highly experimental in a number of critical respects, and labor demands tend to be volatile, owing to rapid changes in production methods, the division of labor, and product configurations. Occupational categories are rapidly invented and just as rapidly rendered obsolete; shifting competitive positions continually undermine profits and wage schedules. Firms need to respond with great agility to changing circumstances. Similar conditions hold for restructuring industries: they need to experiment with new methods in an uncertain, if promising, environment.

In long-established industrial communities, the traditional and institutionalized procedures for stabilizing production activities, employment relations, and the practices of local government tend to inhibit experimentation. Established rights and entitlements for workers and employers, which frequently develop in older industries as solutions to political conflict, collide with the fluidity that is imperative in rapidly developing industries. Moreover, the particular solutions achieved in regions dominated by older industries - involving such matters as technologies, work rules, labor market practices - are likely to be substantively irrelevant to, or in conflict with, the needs of newly developing industries. Thus, the dynamic of capitalist industrialization depends upon the firm's ability to effect changes in political and social relations at the point of production and in the surrounding community. Industries infrequently relocate their centers to reconstitute employment relations alone, but they do frequently move to take advantage of important new opportunities for growth, and the strategic application of labor is deeply imbricated in those opportunities. While such changes may be effected within existing production regions, they are more frequently achieved by redeploying capital investments to relatively undeveloped areas.

Internal solutions to the problems of new industrial development focus on reconstituting the social and political aspects of the employment relation *in situ*. For example, employers may prefer immigrant workers to native-born residents or target first-time female entrants into the laborforce (Nelson, 1986; Sassen, 1988). With older industries, or when significant skills are required, employers may make special arrangements with selected fractions of the laborforce, such as workers with seniority, in an effort to separate them politically from other workers in the region. Local employment relations, then, may be reshaped by developing new forms of labor market segmentation and reconstituting the regional labor supply to fit these segments.

The internal solution has characterized New England in recent years. After a protracted period of disinvestment, from the end of World War II until the beginning of the economic crisis in the 1970s, the New England economy began to grow again even as other regions of the United States slipped into decline. During the period of disinvestment, New England's "labor climate" was powerfully reshaped: the membership rolls and political influence of unions were reduced through unemployment, central city political machines deferred to more conservative suburban governments at the metropolitan scale, and the government began vigorous promotion of economic development (Harrison, 1984)[16]. Meanwhile, New England's educational institutions survived and continued to generate a highly educated scientific and technical workforce, and - critically - to draw federal government funds for both military and civilian research and development.

Internal solutions are inherently problematic. The slate of local history can never be wiped entirely clean. Moreover, restructuring the economic and social bases of an industrialized region is bound to be a slow and costly process, with many setbacks along the way. Where time is not available for the internal solution, space may have to take its place in breaking labor patterns. Therefore, industries frequently opt for radical locational shifts to places where they can create new employment relations and labor politics (Clark, 1981). In this external solution, new communities of firms and workers are constructed on fresh terrain, usually in regions that have not hitherto had an important presence in the industrial economy (cf. Walker, 1981; Young, 1986). Workers must in effect chase after these industries and build territorial social solidarity anew.

The attraction of regions where the external solution is effected has frequently been characterized in terms of high "quality of life" indices (Perloff et al., 1960; Malecki, 1986; Hall et al., 1987). Quality of life is not a transhistorical constant or a universal category, however, but a politically constructed reality; what represents the good life in, say, *fin-de-siècle* Paris bears little resemblance to that in the contemporary US Sunbelt. Quality of life is not a preexisting condition, but a set of social and political attributes crucial to producers - ideologically defined in terms of the consumer - that are created as industries grow. In high-technology industries, for instance, the term usually refers to beneficial arrangements for the key segment of technical-scientific workers, to whom it connotes spacious homes, low-density communities, highly-privatized family life, and abundant recreational resources. For capitalists, on the other hand, quality environments are frequently associated with a "good business climate" - in particular, favorable local tax arrangements, absence of significant labor union activity, and freedom to develop production and labor markets as they see fit. Thus, when a firm moves its plant to a "greenfield" site, it is not showing disregard for the importance of labor in choosing a location.

[16] Many central cities, such as San Francisco, have attempted a similar reconstruction of their labor and employment base (Mollenkopf, 1983; Hartman, 1984).

It is clear that a rising tide of unionization and workers' consciousness in the large industrial cities of the US manufacturing belt laid the groundwork for the industrial restructuring and decentralization that has occurred on a massive scale since the 1950s (Davis, 1986). The highest rates of manufacturing employment growth in the late 1960s and 1970s came about in those areas with the lowest rates of unionization or the lowest incidences of struggle over the employment relations (Korpi and Shalev, 1979, 1980; Peet, 1983, 1984; Therborn, 1984). In turn, this recent socio-spatial constitution of industry and employment has dramatically disorganized the working class and the union movement.

Many new industry growth centers have taken on qualitatively new forms of socio-spatial reproduction. Their very geographical structure both expresses the diminished power of working class and communitarian political movements, and imposes barriers to their resurgence. The dense highly urbanized clusters of workers so characteristic of the older industrial cities are largely absent. Sunbelt cities are characteristically suburban in form, and can be very isolating to individual workers, especially women with heavy household responsibilities and long commutes (Walker, 1977; Rose, 1981). The new immigrants who have poured into US growth centers from Asia are often politically conservative, while Latin Americans have in many cases immigrated illegally, so their social position is especially precarious. Moreover, growth centers in the Sunbelt are typically fragmented into numerous municipalities, which severely inhibits the possibility of replicating big city political machines dominated by working class and ethnic groups.

Nonetheless, the Sunbelt growth centers, like their forebears in Minneapolis, Toledo, and Lille, will not remain smoothly-operating engines of industrialization forever A new process of historical development has been ignited that will eventually lead to organized workers' and citizens' responses (Davis, 1986). It is equally likely that these processes of historical development will offer qualitatively new forms of affective and political life.

6.4.4 *The politics of geographical shifts: growth peripheries*

Growth peripheries are not undifferentiated with respect to labor. As they decentralize, industries still use space to articulate their particular labor demands with supplies set in distinct local cultures. For example, the way electronics companies treat women assemblers in Malaysia, supplying religious shrines or free drugs, would not be possible with (or acceptable to) workers in California (Ong, 1987). Similarly, Japanese auto makers, in building up "Kanban Alley" in Ohio and Tennessee are implanting a set of labor relations that is a compromise between practices at home and the characteristics of US workers and labor market institutions (*Business Week*, 1986; Mai et al., 1988) The result is a complex pattern of industrial centers and peripheries, each with its own characteristic labor supplies, labor demands, and resulting employment relations.

Geographic dispersal of plants often becomes crucial for industries as growth slows – more so if rising worker militancy cuts into their profit shares. Nevertheless, firms cannot effectively shift out of the center unless they free themselves from dependence on the core agglomeration, and such freedom is mediated by technological and organizational conditions, especially those that streamline external transactions and deskill labor processes. To the extent that production can be transformed, and employment relations broken, plants will typically be relocated to peripheral regions in both developed and Third World countries, where there are cheap and captive supplies of labor. Technological change, work reorganization, new laborforces, and locational dispersal may thus move together, rather than any one element following from, or substituting for, another.[17]

The attempt to reconstitute employment relations through plant dispersal can backfire, however. The US auto industry's "southern strategy" of the 1960s and 1970s, for example, did not, as hoped, lower costs. The companies thought they could transplant standardized assembly line technology to cheaper labor supplies but southern workers, for reasons of place and race, had little allegiance either to management or the United Auto Workers, and the employment relations needed to secure productivity comparable to northern plants did not come into being (Clark, 1989). Furthermore, no substantial technological advances had been developed to meet increasing Japanese competition. As a result most southern plants were closed. This experience contrasts sharply with the current success of Ford's new Hermosillo (Mexico) plant or the GM-Toyota plant in Fremont (California) which incorporates production advances along with new workforces and relations.

6.5 Conclusion: the inconstant spatial division of industrial labor

Because labor demand and supply remain differentiated across industries and places, geographical industrialization generates a spatial division of labor. As noted, this division is much richer and more complex than portrayed by hierarchical models; it may be more accurately described as a "mosaic of unevenness." This unevenness is a predictable result of capitalist development given the divergent labor demands of industries and the differentiated social reproduction in territorial communities. Because labor is fundamentally different from other locational factors, the uneven mosaic of industries and places is a necessary condition for, and consequence of, capital accumulation. Employment must allow for the

[17] In contrast to Harvey's (1982) theory of the spatial fix substituting for lack of technical change or the product cycle theory of mechanization-driven dispersal of plants.

mutual participation of classes in production, while at the same time preventing workers from using their power to threaten capitalist reproduction. Stable solutions to this dilemma are only temporarily possible; they cannot be maintained forever: changes in technology, organization, or worker militancy can always upset them. Indeed, stable solutions can become rigid barriers to the competitive flexibility of capital, which must periodically introduce technical innovations or reorganize production. Employment - like capital itself - must forever be in flux, and location is an essential means of shaping and reshaping employment relations.

7 The Process of Territorial Development

7.0 Industrialization and regional growth

Industrial location theory and regional growth theory have too often been considered as isolated entities, despite recognition of their essential unity (Isard, 1956, 1969; Pred, 1966, 1977). The concept of "geographical industrialization" captures that unity from the side of industry growth and location, but we must still inquire into the broader process of territorial development stimulated by capitalist industrialization. The spatial expansion, integration, and division of growing industries and industrial ensembles provide the main shape of territorial development in advanced capitalist economies. Dramatic, but highly specific, transformations of the macroeconomic landscape are effected by particular industries and groups of industries as they undergo rapid growth, reorganization, and technological change. New units are added to the system, and are situated in new ways with respect to existing economic centers, and the space economy is reshuffled over time. This process is especially marked with the appearance of new dominant industrial ensembles.

We prefer the term "territory" to "region." Territory is less theory-laden and more open to fresh connotation; it can refer to any geographical scale, as it denotes functional interaction rather than bounded spaces; a fabric of related places with some coherent linkages may constitute the territory of an industry, or a "territorial complex." The concept of region suffers from two being unduly identified with subnational regions, whereas the developmental processes we are concerned with take place at the subnational, national and international scales at once. The concept is further handicapped by a long tradition of treating areas as self-evident units, such as the Mississippi River basin, the state of Georgia, or the northeastern manufacturing belt. It is, moreover, often taken to be a natural rather than a socially constructed and reconstructed fabric, yet territorial interaction occurs through a wide variety of social processes, from measurable flows of capital, commodities, and labor to less tangible place-related social power relations, traditions, and ideologies.

We consider first how new growth centers expand an existing space economy. Next, we look at the way existing cities and territories are redeveloped and restructured. Finally, we generalize the model of expansion and reconstitution of capitalist territory in terms of broad waves of growth resting on new industrial ensembles and regimes of accumulation.

7.1 The expansion of new territorial complexes

We begin by examining territorial extension by a new industry. The industry creates new growth centers, becomes integrated with established territories by means of inter-regional trade, and develops further by dispersing its growth peripheries in and around the wider territorial system.

7.1.1 Extension by propulsive industries

As before, our story begins when a new commodity-producing sector arises, either due to some genuinely unique invention or to a breaking off from an existing product line. This sector is superprofitable, thus better able to leapfrog to any of a wide variety of places. These locales are unlikely to be entirely without experience of industrialization and may have contributed to the industry's shift by generating technological innovations and pioneering firms of their own. Moreover, new or renewed industries have strong motivations for shifting out of well-developed industrial territories, for example, to avoid a mature, unionized, politicized working class. These initial moves can lead to spatially concentrated growth via large scale factory production and/or collective specialization, diversification, and externalization of production in a disintegrated complex.

There is no *a priori* reason why new growth centers should locate in the largest cities of the urban hierarchy, given the window of locational opportunity afforded initial firms. Major new industrial ensembles frequently favor second- and third-tier places, or areas just on the edge of developed industrial regions: in the United States early in this century, electrical machinery in the Hudson and Mohawk Valleys, furniture in Grand Rapids, and rubber in Akron; in contemporary Europe, consumer electronics in southeast Wales, aerospace in Toulouse, and revitalized craft industry in central Italy. These agglomerations, by inducing territorial development, effectively expand the intensively-developed portions of the national space economy. The US electronics industries are striking in their avoidance of the industrial Midwest (Scott and Storper, 1987). California's Santa Clara Valley is the largest and most important of the electronics growth areas because it is the undisputed center of the *primum mobile* of the entire ensemble, i.e., the semiconductor; it embraces personal computers, mainframes, missile guidance systems and medical-scientific instruments as well. Additional locales focus on specialized uses of electronics: personal computers and biomedical instruments in Orange County; aerospace and military electronics in Los Angeles; aerospace and electronic instruments in Dallas; mini-computers in Boston; super-computers in Minneapolis.

Of course, territorial extension unfolds in different fashions in different countries owing to substantial variations among national political economies, as well as areas and resource bases. Not only is the United States bigger than European countries, but land is easier to purchase and develop; the United States therefore offers a more extensive periphery, particularly in the South and West, and city-building offers abundant profit opportunities for a wide variety of capitalists (Norton, 1986; Logan and Molotch, 1986). The labor markets of the United States and Europe are markedly different as well, with greater labor market differentiation, higher levels of individual mobility over longer distances, and more mass immigration to bolster labor supplies in the United States (Thernstrom, 1964; Thomas, 1973). Even in the face of these powerful differences in the economic payoffs to locational shifts, as new ensembles of production sectors arise, they frequently locate away from areas industrialized by previously dominant sectoral ensembles. In Britain today, the growing electronics industry is largely in Scotland and in the M4 corridor across southern England, not in the English Midlands, once the country's industrial heartland. Likewise in France, high-technology sectors are to be found in such "sunbelt" locations as Toulouse, Grenoble or Montpellier, rather than the Nord/Pas de Calais or Alsace-Lorraine; and, for the German case, in Bavaria and Baden-Württemberg, not in the Rhine-Ruhr.

The main agent of territorial expansion of the industrial economy is the growth center of the emerging industrial ensemble. Following the pioneering work of Perroux (1950), Boudeville (1966), and Hirschman (1958), we emphasize the role of the propulsive industry in the formation of a growth pole. But there are critical differences. We earlier admonished growth pole theories for their crude understanding of the circumstances under which an industrial growth pole could be turned into a territorial growth center, since they failed to account for the intertwined roles of the technical and social divisions of labor which underlie linkage patterns. A second major gap in growth pole literature comes from its lack of concern with the relationships of territorial growth centers to the national or international space economy as a whole, other than policy prescriptions for locating growth centers in backward regions (Richardson, 1973; an exception is Alonso and Medrich, 1978).

7.1.2 Trade relations between old and new territories

A new growth center must be spatially integrated into existing core territories by means of inter-regional trade. Because extensive growth poles develop dense exchange networks with previously industrialized regions, many researchers have mistakenly identified markets in older areas as the motor of regional development. In models of export-led growth, for instance, aspiring peripheral areas are pulled upward by vigorous demand for their output from core territories. As trade continues, the regions mutually reinforce each other's growth economies of regional specialization (North, 1955). Export-led growth has been perhaps the most widely followed theory of development in the postwar world, with Japan as the prime exemplar. The basic problem with the theory, however – as Tiebout (1954) pointed out long ago – is that trade is defined as intra- or inter-regional depending on how one draws regional boundaries. In this vein, export-base theory has been questioned by those who hold to the notion of development via trade and division of labor, but have found historical evidence in Pennsylvania and the Midwest for strong local, or intra-regional, trade patterns between city and farm economies or among territorial systems of cities and towns (Lindstrom, 1978; Meyer, 1983). Lest this throw us back into the closed and static world of central place models, however, there is also compelling evidence that large United States cities have, from very early on, traded more with each other than with their own hinterlands (Pred, 1980).

The apparent paradox of internal versus external markets cannot be resolved by dismissing nineteenth-century experience as irrelevant owing to poor transport infrastructure, since the developed countries were rather well integrated by canal and coastal shipping by mid-century, and railroads came soon thereafter (Fishlow, 1965; Vance, 1986). The source of the dilemma is, rather, an inappropriate focus on trade as the motive force behind growth and inattention to the ways in which the productive forces of industries and territories develop. Demand is, of course, a necessary condition for industrial profits and expansion. But the sufficient condition for industrial and territorial growth is expanding productivity through dynamic economies of mechanization, division of labor, learning, and so forth.

Any dynamic industry must, by virtue of scale economies, ultimately develop into an exporter, and most do so rather quickly as its production outruns local markets. The localization and agglomeration of industries is, by definition, only possible because they can satisfy demand over large market areas from those sites. Export-base theory picks up the story at this point and misreads it as the helping hand of the consumer region reaching down to pull up the producing region. There are, to be sure, cases that partly fit this model: English demand for cotton and grain was instrumental to the size and speed of expansion in the American cotton and wheat industries in the nineteenth century (North, 1961). The cause of such fast-growing consumption in England was, nevertheless, not inter-regional specialization but the Industrial Revolution. Furthermore such industrial inventions as the gin and the reaper were crucial to the ability of the US exporting regions to meet demand.

The most fundamental aspect of nineteenth century US development, however, was the way manufacturing industries such as meatpacking, canning, farm implements and textiles, located in towns and cities, were linked to dynamic and rich agricultural economies in a mutually reinforcing way. This took place largely *within* such growing territories as Pennsylvania, the Midwest or California (Lindstrom, 1978; Post, 1982). These industries became growth poles because of their inner dynamism as well as dynamic interaction with complementary sectors; farmers were mechanizing, breeding, irrigating and rationalizing field practices; Midwestern packers were inventing the disassembly line, the stockyards and refrigerated transport; textiles in Philadelphia were developing an immense disintegrated complex of great flexibility and productivity (Danhof, 1969; Pudup, 1983; Scranton, 1983).[1]

Gradually, a process of "import substitution" takes place that alters the position of the growth center in the space economy. Regional import substitution occurs because, as growth centers draw in population and industry, local markets become a magnet for industries based in older territories (Hansen, 1982; Norton, 1986). Activities with low optimal scales of operation or extremely high unit transport costs, such as bottling plants or bakeries, move in to serve local markets, as described in traditional central place or Weberian location theory (Perloff and Dodds, 1963). Some of the activities thus attracted will be the branch plants or growth peripheries of industries centered elsewhere, such as the meatpackers drawn to Vernon (near Los Angeles) or the New York investment banks and brokerages moving into booming Newport Beach (Orange County, California) and these may, in turn, attract or stimulate the growth of additional suppliers and intermediaries.[2]

[1] Patterns of trade depend on the industries involved and their development, of course. Some will be exporters from the beginning – movies were never a local good – while some will begin from narrowly bounded markets, as did wines in California or specialized Computer-Aided Design equipment in Silicon Valley. Some will rely more heavily on imported equipment (and skilled workers) in their early phases, while others will have to create more of their needs from whole cloth.

[2] Such industrial urbanization is characterized by multiplier effects, which have captivated cumulative causation and export-base theorists influenced by Keynesian economics. The multiplier, it should be pointed out, consists of more than the circulation and recirculation of money; it involves "secondary" or "local serving" activities, in the language of export-base theory. These secondary industries are not simply passive conduits of trade, moreover; they contribute additional surplus value and investment to the metropolitan complex, making it grow beyond the limits of the primary or export-base industry. The construction sector in fast-growing places, for example, may employ over 20 percent of the local workforce (Kuznets, 1966). Actually, many such firms or industries are wrongly labelled local-serving – as if they were all barbershops – when they are, in fact, part of the disintegrated production complex that generates certain export products at its apex. Furthermore, such industries, even if derivative at first, may well build up their practical competence and innovate in such a way as to become competitive and growing activities in their own right.

In short, as new industrial complexes are added to the territory of the national or world economy, they tilt the spatial structure of effective demand, altering the locational choices of some industries and skewing the overall locational configuration. At the same time, export targeting, branch plants, and takeovers from outside will capture or eliminate some local industries, as occurred when the Harriman railroad empire moved west to absorb the once dynamic Southern Pacific company in 1900. Secondary industrialization intensifies inter-regional linkages, but in an uneven fashion that frees the growing territory of interregional dependence in some ways while tying it more closely to older centers in others (Loertscher and Wolter, 1980; Caves, 1981; Ethier, 1982). California, for instance, has teetered between bursts of independent industrialization and invasion by imports and branch plants: while gold, mining equipment and oil have been successful export industries, locomotives, explosives, steel and vehicles were subordinated to or conquered by dominant centers elsewhere.

Ironically, then, in successfully developing places import substitution generally takes place *after* dynamic growth sectors have been established: the market largely follows industrialization rather than preceding it. The crucial issue is therefore not import or export *per se*, but the ability to generate or attract dynamic growth sectors that generate a virtuous circle of increasing productivity, income, and consumption.

7.1.3 Expansion of growth peripheries

Even as the growth center increases in size, output, and organizational complexity, countertendencies are often at work, and portions of its propulsive industries may begin to disperse into new growth peripheries. Such dispersal can take several forms, as we have indicated: it can be pulled by market penetration, pushed by labor militancy, propelled by technological standardization, facilitated by large-scale factory integration, or forced by competitive pressure. But in the context of rapid growth, the centrifugal force of an industry will be powerful regardless of the exact shape it takes.

Let us introduce three subsidiary processes of dispersal, involving the amplification of growth peripheries by secondary development. One is the stimulation of complexes of supplier industries in the hinterlands of the growth center of a propulsive industry. A classic example is the way rubber, glass, and machine tool industry complexes served the dominant auto industry of Detroit and southern Michigan from subcenters in Akron, Toledo, Dayton, and so on, thereby creating an immense territorial cluster based on the ensemble of auto-related industries. A second is the formation, around dispersed plants, of secondary complexes devoted to producing inputs, but this time within the propulsive industry itself. These may be wholly-owned corporate subsidiaries, independent "merchant" firms, or take a variety of intermediate organizational forms usually termed subcontracting. An increase in the scale and specialization of production units can make possible the formation of specialized complexes that cross-ship to each other (Ethier, 1982). These long-distance relationships are frequently of an intra-firm nature, as in IBM's system of integrating its European plants and subcontractors (Bakis, 1980), owing to economies of internalization in managing far-flung systems and to keep a handle on proprietary technologies. A third path of development of growth peripheries is the transformation of a low-cost branch plant or assembly operation into a more integrated center of production for expanding regional markets, as in the case of Japanese and American electronics plants in Singapore and Malaysia (Scott, 1987).

This decentralization generates an intra-industry spatial division of social labor. The spatial division of labor of the propulsive industry rarely duplicates patterns established by the industries that came before, as theories of hierarchical order and stability would lead us to expect. Decentralization is not an even, predictable process, but is highly selective in time and place among the non-metropolitan places, smaller cities and greenfield sites available (Clark et al., 1986). Decentralization is the strategic and selective creation of a particular spatial division of labor in each industrial ensemble, given its historical and technological conditions. The combination of two distinct subprocesses – the formation of territorial growth centers by leapfrogging and the formation of growth peripheries by decentralization – allows newly dominant industrial ensembles to bring a shift in the macro-regional pattern of economic growth.

7.2 The redevelopment of old cores

Older cities and regions are subject to many of the same processes as new centers: renewed growth in established sectors, the creation of new propulsive industries, elaboration of the division of labor, reorganization and new patterns of linkage, restructured labor relations, and so on. There are two especially important causes of such redevelopment: reinvigoration by the new industrial ensemble through a process of "reverse diffusion," and expansion of portions of the division of labor identified with the highly agglomerated office-based "service" industries.

7.2.1 Reverse diffusion from propulsive ensembles

One of the principal ways that innovations in technology, organization, and labor relations affect regional development patterns is by reorganizing and redefining older industries. Industries that do not form part of the new, dominant ensemble may be dramatically reinvigorated by innovations deriving from that ensemble, and the resulting spatial changes may be as dramatic as those associated more directly with the behavior of the new dominant industries.

Consider a few major lines of transmission of technological change. First, inter-industrial applications of the outputs of new capital goods industries may be promoted through learning, and thus transform the production processes of the adopting industries (as with steam engines in the nineteenth century and microelectronic controls today). The effects of such applications are often quite varied and elaborate. In the simplest case, an innovation is applied to the production process of an existing factory, and lowers production costs or raises product quality. It may also require altering the design of an existing product. If this changes the input–output structure sufficiently, the search for more suitable factor supplies may lead to locational change, or the need to introduce new methods of automation may lead management to relocate in search of a new laborforce.

This dramatically oversimplifies the potential effects of innovations on older industries, because it ignores the changes innovations may induce in the industry's social division of labor. For example, a major renewal of fixed capital stock – where incremental technical changes are pulled together into a new conception of the production process – often involves investing in new units of production. The old division of labor among factories may give way, with some phases of the labor process separated into completely independent units while other, separated phases are recombined. As a result, new industrial complexes may appear or old ones may break apart. Consider the application of a new base technology – microelectronics – to an existing industry, automobiles. General Motors is currently experimenting with flexible automation. It attempts to cut costs by making it possible, within limits, to shift between product lines by organizing factories to produce groups of cars rather than any one specific model. This strategy is designed to improve capacity utilization over time. As a complement, GM is developing just-in-time delivery systems for parts and in-factory programming. This system suggests the re-formation of a regional industrial complex as the core of the company's production system: short supply lines, precise delivery times for parts and components, and frequent order changes require spatial clustering of subsidiary units and subcontractors (Ikeda, 1979; Estall, 1985; Schoenberger, 1987a).[3]

[3] Ford Motor Co. seems to have gone a different route, adopting the "world car" form of organization, in which production is vertically integrated within the company, pushed to maximum economies of scale, and integrated via long-distance, highly-standardized flows of material and outputs. Cost reductions are achieved via economies of scale, orienting the system to a global, or at least multi-continental, market, and using the extensive locational capabilities of a vertically-integrated production system to locate production units in areas of low labor cost – and to force wages down in high-wage areas by making examples of the plants located in low-wage regions. In essence, Ford has pushed the production system it developed for use in the United States in the 1950s and 1960s to a higher spatial and output scale. The local industrial complex is, effectively, no more. The GM system, on the other

As a result of the formation of these new industrial complexes – the division and reintegration of production in new ways – a new set of spatial production relations may come to characterize the spatial division of labor and to change the relationships between places in the space economy. The inter-scalar relations of the space economy (between different cities, regions and nations) are not inert features of some spatial geometry, nor can any set of them be assumed to be historically permanent.

New sets of inter-industrial divisions and complementarities may also be forced by the introduction of new final outputs. New consumption goods industries, for example, may have an impact well beyond their own narrow borders, because once their products are inserted into daily life they alter other social practices and demand complementary goods and infrastructures, while eclipsing the need for existing goods. The most obvious example is the appearance of the automobile which has had enormous effects on patterns of consumption, and has largely replaced intra-urban rail travel.

Changes have effects well beyond those just discussed, however, for core technologies have different externalities and diseconomies that promote (or prevent) combinations at the product or process level. For example, a given process may be adapted to a variety of product outputs. As firms developed new techniques in the nineteenth century for the manufacture of firearms, one firm found these techniques applicable to typewriters. As a result, Remington came to specialize in both from a single regional location (Rosenberg, 1972). The new process encouraged a greater scope of outputs, and so these vertical technical externalities encouraged inter-industrial combination.

On the other hand, a new component that fits into a core technical system may well become a shared input for a number of output sectors. The input may best be manufactured by firms that specialize in it, reflecting vertical technical diseconomies. In this case, technology transmission usually takes place as a consequence of the use of these discrete product outputs in different, unique combinations. Yet production of the components, however modified, may remain spatially separate, embedded in different territorial complexes. This can be true, even where firms in new industrial ensembles are purchased by firms in older ensembles. For example, General Motors recently acquired Electronic Data Systems, based in Dallas, and is presumably using EDS to customize electronic components; but the actual merging of EDS's activities with GM's central

---

hand, stresses programmable machine systems, decentralized inventory control, and close-knit communications, rather than central guidance of worldwide operations, except at the level of corporate planning. The economic rationale is to minimize the size of production runs, exert downward pressure on components prices, shed inventory and overhead, and maximize system flexibility. Recall that this follows on GM's unsuccessful "Southern strategy" of the recent past, noted previously.

supplier complexes is not yet possible because the base technology is still discrete. A new base technology has led to a new set of components for automobiles, yet those components are most efficiently produced in industrial complexes linked to their own technologies, not to places of final product use. This might very well change, with further development of the process technologies for the electronic components.

Obviously the possibilities for new inter-industrial and inter-firm technological linkages are immense and diverse. But the transmission of technical change cannot be reduced to transmission of techniques embodied in products and machines, for production involves much more than hardware. New dominant production ensembles often bring with them new forms of the social organization of production: new social divisions of labor between firms and new types of firms; new shop floor layouts, ways of working, and wage systems; and new styles of management. Many of the innovations in the organization of production arise alongside new technologies and so provide organizational and management models for older industries which can be implemented independent of any technological hardware (Reich, 1986; Piore, 1987). For example, both Fiat and Olivetti in Italy experimented with advanced automation on their shop floors in the early 1980s, only to discover that most of the productivity gains they sought could be achieved by reorganizing the shop floor, supplier-buyer relationships, and wage systems – without the newer technologies (Locke, 1987). The new wave of Japanese-style industrial organization may be less due to technology than to rethinking organizational aspects of production (Schonberger, 1982; Sayer, 1986b). Chandler's (1962, 1977) classic analysis of the modern corporation suggests that the success of Fordist assembly-line production was greatly advanced by Alfred Sloan's organizational innovations at General Motors. Thus, when we refer to major periods of rapid inter-industrial transmission of innovation, we must think beyond machines to organizations and ideas.

The conception of inter-sectoral transfer of innovation advanced here may be contrasted with the notion of innovation diffusion which dominates the literature. After a long history in cultural geography, diffusion theory entered economic geography through the work of Hagerstrand (1953).[4] In economics it developed independently out of the work of Griliches (1957), but soon lost its geographical content (Feller, 1975). Diffusion theory ordinarily treats innovations as new ideas which move principally through information flow. Users are assumed to be ignorant of the innovation until news of it arrives, but able to employ it once it is presented to them. This is known as the "adoption" perspective (see, e.g., Rogers and Schoemaker, 1971). Diffusion is therefore only limited by barriers to information circulation, such as simple distance (neighborhood effects), and by the potential adopters' resistance to change. The diffusion of any

[4] For reviews see Pred, 1974; Blaut, 1977; Brown, 1980.

innovation is said to follow an S-shaped curve as a new idea is tried out and barriers are gradually overcome until a bandwagon effect takes place; finally, as the potential field of users is exhausted, new adoptions diminish in number.

Diffusion theory is closely bound up with the activity of agricultural extension agents among prosperous farmers in North America. In this farm service model, innovations arise in model research and development centers and are brought to modernizing farmers by outside experts; the landscape is filled with relatively homogeneous producers, all with access to the means of production (Griliches, 1957). Urban applications of diffusion theory rely on the same basic vision: innovations originate at certain nodal points and diffuse through a homogeneous field of users. But the model is amplified in two ways. First, urban-industrial centers are portrayed as "seedbeds" of innovations, "incubators" of new firms, and "hotbeds" of spinoff activities. As Hoover put it, established industrial centers serve as "germinating grounds" for new industries because of the availability of "versatile labor and venturesome capital" (1948, pp. 174–5). Second, innovations are said to filter down a hierarchically structured urban system (Thompson, 1968; Hudson, 1969). Pred (1974) has modified this by allowing for diffusion up and across the fabric of the urban system, but still grants the largest centers the power to capture innovations because they dominate flows of information and economic activity, which leads to the same outcome: stability in the rank-size ordering of cities.

The diffusion literature has been preoccupied with the diffusion of innovations in space rather than the effects of adoption on locational behavior, spatial production relations, and patterns of regional development. As a result of the split between theories of location and diffusion, almost no work was done on the locational implications of the adoption of innovations by manufacturing (Feller, 1975). Yet historians of technology have demonstrated that production innovations move through economic networks of shared inputs and product groups. Nothing leads us to expect these networks to take on a universal and geometric spatial order, especially not a neat hierarchy. Indeed, we would argue that the term "reverse diffusion" is more appropriate than hierarchical diffusion, in two senses. First, technological and organizational innovations are likely to emerge from new sectors and dominant production ensembles and to find their way back to existing industries. Second, the geographical centers of new ensembles are often on the margins of developed industrial space. If diffusion has a tendency to move in any spatial direction, its initial wave is then both up the urban hierarchy and inward to older industrial centers and regions. Ultimately, however, the most important effects of these innovations may be when adopting them leads to a reforming of location patterns and of spatial production relations. This may, in turn, cause the redevelopment of both old and new industrial regions in accordance with the formation and reformation of the social division of labor and its associated linkage structures and factor demand patterns. We may thus conjecture that the transformation of US industrial space by the high-technology industrial ensemble has by no means run its full course: it may bring further interesting and subtle changes in "mature" industries such as automobiles, tires, and machinery.

7.2.2 Office-based industries and the renewal of old centers

Urban and regional instability is caused by spatially differentiated industrial growth and decline, and results in a reshuffling of positions in the urban hierarchy. This view of disequilibrium territorial industrialization is not entirely incompatible with circular and cumulative causation models based in urban economies of scale and disequilibrating growth. Nonetheless, cities and regions thrive not on agglomeration in the abstract, but on the fortunes of their constituent industries, which must be renewed or replaced over time. Conversely, the decline of agglomerations is not solely due to diseconomies in the abstract, but to new geographical opportunities emerging in production. Cities are forever subject to changing "urban fortunes" (Logan and Molotch, 1986).

Deindustrialization has eliminated innumerable small cities from prominence in the urban system. Economies of agglomeration are often said to protect large cities against such a fate. But the process of agglomeration is selective. It pertains, in the first instance, to the divisions of labor within technically-related industries, as argued in chapter 5, and only subsequently to miscellaneous secondary industries, as indicated in this chapter. Large metropolises, such as London, New York and Paris, are not just very large agglomerations. Rather, they are spatial aggregations of multiple industrial complexes, in that the metropolis or region serves as a true center for several distinct industries. As we have noted, Los Angeles is a center for movies, aircraft, aerospace electronics, furniture, and garment production, each with a spatially discrete submetropolitan field of operation. Although a diversity of industrial activities can act as a buffer against economic fluctuations in any one sector, even the largest cities can suffer from the decline of their propulsive industrial agglomerations – as in Detroit, Pittsburgh, and Liverpool, for example. The same applies to entire regions. New England, undisputed leader of American industry in the nineteenth century, could not overcome the loss of its pivotal shoe and textile industries until late in the twentieth century, when electronics growth finally reversed a long decline.

Even in diversified regional economies continued growth depends on adding new or renewed localized sectors as old ones decline, as the recent histories of New York City or Los Angeles illustrate. The decline of some sectors is often hidden by the vigorous expansion of others – even to the point of taking over recently shut plants, as in Northrup's use of Ford's former Pico Rivera, California, factory to manufacture Stealth bombers. Most frequently new growth industries attach themselves to the peripheries of existing metropolitan areas, as happened with petroleum refining or canning in the San Francisco Bay Area at the turn of the century. Nonetheless, some new fast-growing sectors spring up near the heart of the metropolis, as, for example, the recording industry of Hollywood, the shipbuilding industry of the San Francisco Bay Area during the two World Wars, or the contemporary biotechnology activities of several big cities such as Minneapolis/St Paul. Nothing in our theory absolutely precludes rising sectors from paying a high price for central city locales, including significant costs of urban renewal.[5] But which sectors might be so inclined? The principal group today are those occupying office-type buildings and widely referred to as "service" activities.

The expanding social division of labor is a principal source of productivity increase in the industrial system. The social division of labor changes because new products are invented, but also because production processes are extended and elaborated, increasing the roundaboutness of industrial activity: important dimensions of this expansion involve pre-production activities such as research, development and design; post-production work such as packaging, selling and advertising; administrative functions including accounting, hiring, training, and planning; and financial activities such as banking, securities trading, and insurance. These portions of the overall social division of labor have come to be called "services." That term can be highly misleading, however, for these kinds of labor have little to do with "serving" anyone directly (i.e., labor-services). Rather, their function is to serve capitalist industrialization by raising labor productivity, multiplying the number of products, circulating commodities faster and more effectively, circulating money and providing credit, and administering an increasingly complex system – all aspects of capitalist economies that go back to the dawn of industrialism (Walker, 1985a). Nonetheless, in both absolute and relative terms, occupations associated with these forms of indirect labor have steadily increased, giving rise to the mistaken idea that a post-industrial, or service, economy now prevails (e.g., Stanback et al., 1983).

Service activities have become closely identified with the transformation and continuing growth of major cities already high in the urban hierarchies of advanced capitalist countries (Noyelle and Stanback, 1984). Some observers point to this to confirm the idea of stable hierarchies of places in which service jobs are somehow superior to ordinary manufacturing occupations (Lipietz, 1980; Massey, 1984). Others have argued that the newer service activities follow a universal spatial development pattern of

[5] Closer inspection of the microgeography of such sites, however, often reveals that they are indeed "peripheral" to centers of growth of competing industries. For example, most shipbuilding in the Bay area in World War II was done in unoccupied portions of Richmond and Sausalito, at the water's edge and on filled land.

"birth" in the largest centers and "diffusion" down the rank-size hierarchy (Daniels, 1979). These notions are appealing in their simplicity and regularity, but are belied by the evidence. A better way to account for the evidence is to show that renewal of older cities involves not services but office-based industries; not corporate or skill hierarchies but agglomeration and localization processes; not initial advantage or seedbeds but selective renewal and decline.

The office-based industries[6] have long been disproportionately clustered in large cities, especially in their central business districts. Merchants and the earliest banks congregated thus in the eighteenth and nineteenth centuries; the first stock markets, commodities trading markets and investment banks did the same in the late nineteenth century, followed by advertising and mortgage banking in the early twentieth century. When corporate headquarters first began to separate from manufacturing facilities around 1900 they, too, were drawn to the city center, and today the swelling office-based producer-services industry is following the same pattern.

All this activity rests on certain basic conditions of producing services. One common thread is that the principal labor process of most office-based industries is information-handling. As a consequence, their labor processes tend to be highly unstandardized, requiring considerable personal knowledge and continuous social interaction within and between workers in interrelated parts of the production system. This, in turn, determines the high level of transactions, the unstructured knowledge on which so much business rests, and the need for continuous information flow. Furthermore, such labor processes lay the foundation for the multiplication of highly disintegrated production units and the development of the office complex itself into a rich nexus of interdependent activities with considerable external economies of information-sharing, close and continuous interaction, and flexible response to changing market and investment conditions (Hoover and Vernon, 1959). That is, two kinds of roundaboutness of production have occurred: between the office sectors and the rest of the economy and within the office sectors themselves.

It was common in the 1970s to hear predictions that improved telecommunications would lead to a decentralization of information-based industries. Yet for the most part this has not happened. On the contrary, an enormous urbanization boom occurred throughout the United States and other countries in the capitalist world in the 1970s and 1980s. The reason can be found in the countervailing force exerted by dynamic external economies in the organization of production in these industries. Moreover, the general ease of modern communication actually facilitates the growth

[6] The term "industry" has long been treated as virtually synonomous with manufacturing. There is no reason for this, as the definition of industry is independent of the material content of either its products or production methods (see chapter 5).

of the functions basic to central city agglomeration, by making possible even larger multinational corporations, bigger banking networks, more international securities transactions. The same technologies also stimulate the invention of any number of new producer services, such as computerized securities trading, and these specialized activities are themselves often carried out through an elaborate social division of labor. Thus, there is no end to the interaction of scale and specialization in the industrial economy, and there is no reason, in principle, to expect an end to agglomerative forces, even with continued improvement in transport and communications technologies. Continual advance in the productivity of social labor as a whole generates new, highly unstandardized branches of production which renew agglomerated production complexes.

Subsectors of the office-based industries are quite markedly localized. Wall Street is the US center for securities transactions, with only minor competition from San Francisco, Los Angeles and Chicago, and bids fair to transcend London in the trans-Atlantic markets. Chicago, on the other hand, is the world capital of commodities trading and futures markets, surpassing even New York and Tokyo in securities futures owing to its specialized work in commodity futures. The San Francisco Bay Area has become the world leader in venture capital, thanks to the particular history of financing electronics, although New York, Chicago and Boston retain secondary pools of money, firms and expertise in the United States (Florida and Kenney, 1988). Similarly, New York City has been the historic focus of the advertising industry in the United States and Hartford, Connecticut, the heart of the US insurance industry.

The recent growth of headquarters, finance and producer services has not benefited all cities in proportion to their hierarchical standing. Much of this office activity has intensified the development of existing centers in the national space economy, but it has been selective. In the United States there has been a dramatic spatial concentration of producer services and corporate headquarters in a few cities (New York, Chicago, San Francisco, Los Angeles), but a slow and steady attrition of those functions away from the central business districts of many others (Toledo, Pittsburgh, Buffalo and St Louis). This shift in the overall pattern of office agglomeration corresponds to a restructuring of office industries, including substantial corporate concentration through mergers and buyouts. A new wave of central city office construction is occurring, as it were, on the ashes of many of the old downtowns. The same differentiation has taken place regionally, as well. In the Midwest, Minneapolis retains a modestly vigorous downtown in large part on the strength of its finance and mercantile functions for the North-central region and for the national grain trade. The central business districts of Cleveland and Detroit, on the other hand, are hollow shells despite the presence of a few large company headquarters.[7]

The decline of office functions in these older cities has an additional cause that lies outside the logic of location for distinct office-based industries, however. Most office clusters arise initially to serve regional production activities; they are part of a territorially-integrated social division of labor between regional manufacturing, mining or agriculture and their administration, financing and merchandising. For example, a number of office complexes arose in American cities, such as Buffalo and Detroit, to service territorial production centers during their periods of rapid industrialization; now many of these industrial centers are in decline, and so are their office clusters. The remaining headquarters of these cities are relics of the manufacturing economies established in the early twentieth century (Perry, 1987; Hill and Feagin, 1987). TRW has roots in a Cleveland machining company but its principal base of operations today is in the Los Angeles aerospace cluster and most of its office functions have been transferred there. Conversely, new growth centers established by new branches of industrial production generate their own firms and associated finance and producer services; several Silicon Valley firms, such as Hewlett-Packard and National Semiconductor, are now among the country's 100 largest industrial companies.[8]

[7] We can see dispersal at work in the office-based industries as well. Standardized back office functions, customarily employing large female laborforces and minimal contact with the central city, have been decentralizing; most have gone to large suburban office centers growing as fast as, or faster than, central business districts, while others have been transferred to such faraway places as the Bahamas or South Dakota for cheaper labor or for mid-American accents among phone operators (Nelson, 1986). At a national scale, the dispersal of midlevel offices of large corporations has not always conformed to the established urban hierarchy, choosing instead to locate in newer midsize regional cities, particularly in the South and West, and often in suburban nodes rather than central cities (Pred, 1977; Daniels, 1982; Noyelle and Stanback, 1984). Even more troubling for hierarchical models is the rather substantial movement of industrial headquarters to the suburbs of large metropolitan areas (Daniels, 1982). This has been explained in terms of factor costs, or commuting distances from executive homes, but these only come into play once the pull of central city agglomeration economies has diminished as corporations internalize office functions. Indeed, the real pillars of central business districts, we would argue, have never been the *industrial* corporations as much as banks, merchants, or producer services – for whom information access and disintegrated production complexes are the most crucial.

[8] Banks have much the same relation to regional economies, which has only been attenuated for a few international giants. For example, San Francisco's Bank of America rose on the strength of the California economy, and its decline is due as much to bad farm, oil and real estate loans there and in the United States as to Third World adventures. The ability of Los Angeles finally to surpass San Francisco to become the second financial city of the United States is testimony to the greater speed of regional industrialization in southern California, despite the historic advantage of San Francisco as a West Coast financial center. The same may be said of the meteoric rise of Japan's great banks in the world system; Tokyo is now the leading banking center in the world, Osaka the fourth.

7.3 Waves of territorial industrialization

We want now to extend the argument about territorial development to embrace very large-scale processes of geographical industrialization in the history of capitalist development. Industrialization over the last two centuries has been structured in a general way by technological ensembles which have far-reaching application throughout the production system. These technologies have provided the framework for long upswings of growth, and the foundation for the shifting fortunes of regions and nations as capitalism has spread around the world.

7.3.1 The evolution of technological ensembles

Technological ensembles may carry forward such a wide range of industries that they define whole periods of industrial progress in a way that transcends the diverse natural trajectories described by individual industries. Epochal technologies do not overtake all industries evenly, however, and even leave some relatively untouched: because of their technological specificity, industries differ widely in their ability to exploit such advances. As a result, it is possible to speak of clusters of leading sectors propelled by the unfolding possibilities of one or more base technologies, as with electronics, chemical, aerospace and a handful of other sectors in the post war era (see figure 7.1). These can be so broad that they embody the dominant technological spirit of an era, as in the commonplace terms "machine age," "Fordism," "automobile age," "electronic age," and so on. Despite the absence of definitive research on the scope of leading edge technologies over the course of capitalist industrialization, it is possible to indicate the lineaments of such technological eras and their key ensembles.

The first set of technologies to burst forth in the Industrial Revolution were spinning and weaving, iron smelting and casting, agricultural rotation and soil improvement, and prime movers for mills (textile, saw, flour, etc.), mines, and railroads (Mantoux, 1961; Landes, 1970). Industrial technology in the second half of the nineteenth century was transformed, above all, by advances in machine-making. This activity became an industry in its own right after breaking off from textiles during the 1840s, and soon everything from boilers to locomotives was being made on an industrial rather than a handicraft basis (Marx, 1867). Machinery-making rested on the ability to transform iron into steel, and progress in metallurgy was represented by carbon-steel and copper smelting, but the decisive role was that of the machine-tool industry, which cut and shaped the metal parts which allowed the fabrication of more accurate, durable and high-speed machines (Rosenberg, 1972). Machining and steel-making together made possible most of the secondary technical clusters that define the

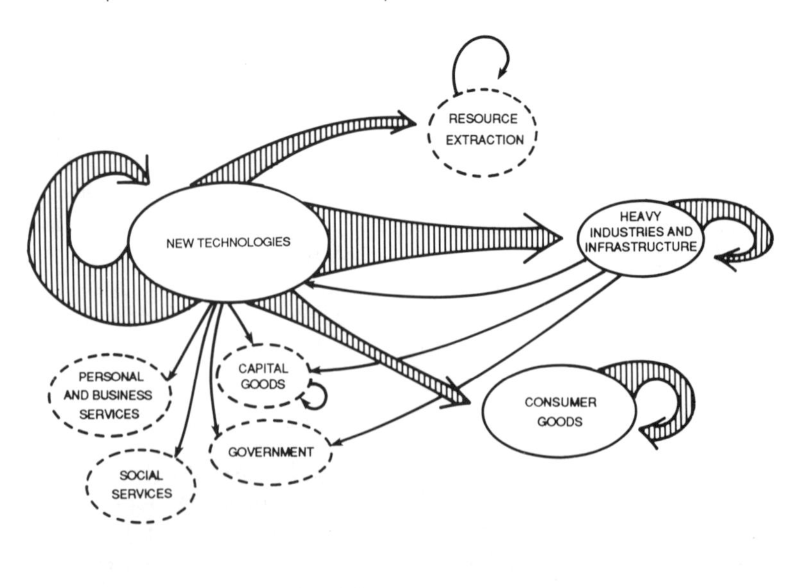

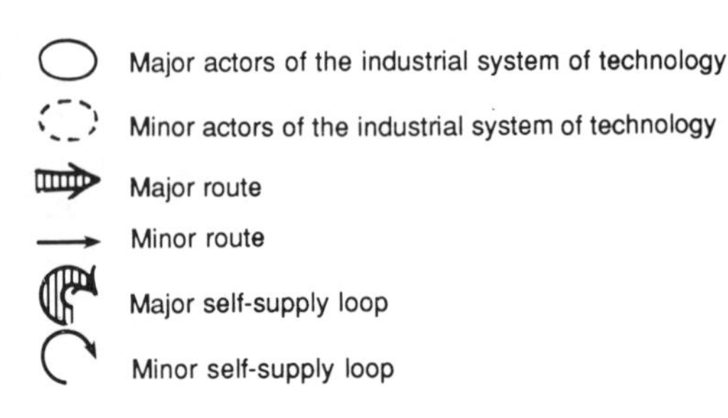

Figure 7.1 *Flows of technology from leading sectors to the rest of the contemporary economy* This diagram portrays the weight of technological innovation and paths of diffusion outward from a handful of key sectors in the contemporary British economy in the late 1970s. The sectors depicted are:

New technologies	aerospace, chemicals, computers, electrical, nuclear, instruments, tankers
Resource extraction	agriculture, fishing, forestry, mining
Heavy industries and infrastructure	materials, energy, communication, storage, transportation, utilities
Consumer goods	appliances, apparel, automobiles, beverages, cosmetics, detergents, food, furniture, leather, rubber, tobacco
Capital goods	construction, machines, production facilities, tools
Personal and business services	entertainment, leisure, personal and consulting services, publishing, recreation, television broadcasting
Social services	education, environment, health, welfare

(After DeBresson and Townsend, 1978)

machine age: the railways and steamers that revolutionized transportation; the farm equipment (steel plows, harvesters, diskers, etc.) that propelled agricultural labor productivity; more powerful and accurate firearms that expanded the art of death-dealing; the sewing machine, which created a true garment industry; metal rollers that made over the grain milling industry; ball bearings and frames for bicycles (the first true consumer durable); canning machinery that made canned food universally available; wood-working machinery that helped make saw-timber and mass-produced houses possible; elevators and steel reinforcing that allowed office buildings to soar ten stories or more.

At the turn of the twentieth century capitalism was swept onto a new plane of industrial achievement with the capture of the genies of electricity, inorganic chemistry, and the internal combustion engine (Schumpeter, 1939; Freeman, 1982). The electric dynamo and motor made possible the trolley car, electric utilities, and household electrification; electric motors lent a new flexibility to machinery, while electric sensors and instruments added new dimensions of control to machine systems (Hirschhorn, 1984). Lamps, heaters and small appliances became important new consumer products; the first electronic devices such as the radio, telephone and victrola gave birth to the recording, broadcasting and communications industries; and electrolytic processing unlocked the secrets of aluminum and metal plating. Electricity also helped launch the modern chemical industries, such as nitrogenous fertilizers and chlorine-based substances; the latter, merged with other breakthroughs in inorganic chemistry, yielded a major bulk chemicals sector; advances in photochemical processing led to a commercial film industry and the first consumer photography, while organic chemistry pushed beyond aniline dyes to the first synthetic fibers and plastics such as rayon, cellophane, and nylon, and to modern pharmaceuticals.

Enter next the automobile, driven by a revolution in mass production and mass consumption: cars and trucks consumed huge quantities of steel plate, machined parts, glass, rubber, wood, and leather, and gave birth to a whole new industry of tire-making; oil refining grew into a first-rank industry on the strength of the demand for gasoline; gasoline and diesel engines transformed shipping, altered railway locomotives, and made the airplane possible. Ford's assembly line spread to other consumer durable industries, such as refrigerators, irons and vacuum cleaners. Finally, a new era of large-scale construction began on highways, dams, factories and skyscrapers, propelled by heavy machinery such as cranes and bulldozers and aided by new materials such as Portland cement, steel girders and asphalt.

Around World War II a new cluster of technologies came into being. The single-wing and jet aircraft became a major civilian transportation mode, and spawned huge airport complexes, while airplanes, helicopters and missiles became the mainstays of the military-industrial complex. Oil refining spawned a petrochemicals complex by generating, besides fuels, a wide array of hydrocarbon distillates suitable for use in everything from solvents to pesticides; most important of these were the feedstocks for plastics such as polyvinylchloride and synthetic fibers such as polyester, from which were derived millions of fabricated consumer and industrial products. Electronics exploded, moving beyond radio to television manufacture and broadcasting, high-fidelity records, and audio and video tapes, and became solid-state and digital, giving rise to the semiconductor and the high-speed computer. Agriculture enjoyed its biggest productivity spurt since the eighteenth century thanks to a cluster of related advances in plant hybrids, fertilizers, irrigation systems, mechanization, and processing.

7.3.2 Technological ensembles and regimes of accumulation

Technological ensembles provide the vital skeleton for aggregate capitalist growth. This is true, first of all, in the narrow sense laid out in the model of growth swings: technology provides the means by which needs are satisfied, the framework in which labor produces surplus value, and the vessel for investment. If technical change unfolds at the right pace, labor productivity rises and products multiply in the market: if the rate of change is more rapid, it allows for a faster expansion of consumption and production and for absorption of new investment; if the rate is slow, it is more quickly outrun by the drive to accumulate, and overinvestment is the result. That model holds on an aggregate level as well as for individual sectors, though it must be modified for the interaction of industries expanding at different rates. Aggregate crises manage to pull down even the strongest sectors at times, while at others a dynamic sector will be able to grow right through a period of general malaise, as computers largely avoided the recession of 1982 in the United States. It is also important to reiterate Schumpeter's idea of gales of creative destruction accompanying waves of innovation in the aggregate growth process. Growth is not a smooth process of adding output and surplus value to that previously accumulated, but a process of constant upheaval due to competition of the old with the new, and amongst those trying to introduce new ways of doing things.

This model of growth is too lean, however. The foundations of capitalist expansion encompass more than technological advance and capital investment. A whole constellation of conditions must obtain for accumulation to proceed: there must be a growing supply of labor-power, wages must be low enough to allow profits but high enough for workers to purchase what has been produced; money supply must keep pace with accumulation, credit mechanisms must exist to bridge gaps in time and space; forms of business organization must be created to manage production competently; a sales apparatus (retail outlets, sales representatives, merchant distributors and the like) must expand with the growing mass of commodities; land must be developed and allocated so that factories, offices and households can be reasonably close, and so forth. Both Marx and Schumpeter had extraordinarily rich conceptions of the actual fabric of capitalist growth. Unfortunately, the subtlety of their work has often been lost by followers who seek to strip the arguments down to essentials; as a result, one-sided notions arise as to the source of growth or crisis, when in fact a whole series of conditions must ordinarily be met if growth is to proceed.[9]

The term "regimes of accumulation" captures in a descriptive manner the particular configurations of capitalist development in different places and different times. It comes from the French "Regulation School" of Marxist economists, which has been particularly attentive to the institutional fabric of a working growth ensemble (Aglietta, 1976; DeVroey, 1984; Lipietz, 1986). "Regulation" refers to maintaining the institutional fabric of growth in a dynamic and contradictory setting through state interventions and class compromises. This group has mostly focused on Fordism as a system of production and regulation. The vastly increased labor productivity achieved through the assembly line method, it is argued, led (after hard experience with the overproduction of the 1920s and stagnation of the 1930s) to an institutionalization of high and rising wages through the collective bargaining system that stabilized accumulation (Burawoy, 1985; Davis, 1986).[10]

All models of growth, no matter how multifaceted, must still assign relative causal weights to key factors in the success and failure of accumulation. We argue that the expanding forces of production are the hard foundation for every regime of accumulation, on which the institutional regulation of the balance between production and consumption is constructed. Without the rising labor productivity unleashed by Fordist mass assembly, as well as other advances such as continuous flow chemical processing, the regulated wage bargain of the postwar era would have been

[9] Their most original followers have also tried to capture the multi-dimensionality of capitalist development (Luxemburg, 1913; Lenin, 1917; Dobb, 1947; Baran and Sweezy, 1966; Mandel, 1975; Aglietta, 1976; Harvey, 1982; Freeman et al., 1982). A common mistake in such discussions is to see structures directly determining outcomes, rather than generating impulses and possibilities that may lead to certain results, under certain conditions. In the living organism, by comparison, DNA can trigger the development of the needed organs for growth, but it cannot assure that all the conditions of growth will be in place: malnutrition can stunt the brain, iodine deficiency can cause the thyroid to fail, and heavy work by girls can cause late onset of menarche. Similarly, capitalist relations of production are a generative structure that may, for example, stimulate credit mechanisms that allow for larger investments in factories, more extensive infrastructure, more continuity in sales, which in turn, advance capital accumulation; yet this credit system can malfunction, setting off a financial crisis that interferes with production (Harvey, 1982).

[10] We should note that there is currently a lively theoretical and empirical controversy over the concepts of "regime of accumulation" and their attendant forms of "regulation."

Table 7.1 Long wave chronologies according to various authors

	1st Kondratieff		*2nd Kondratieff*		*3rd Kondratieff*		*4th Kondratieff*	
	trough	peak	trough	peak	trough	peak	trough	peak
Kondratieff	1790/	1810/17	1844/51	1870/75	1890/96	1914/20		
De Wolff	—	1825	1849/50	1873/74	1896	1913		
Ciriacy-Wantrup	1792	1815	1842	1873	1895	1913		
Schumpeter	1787	1813/14	1842/43	1869/70	1897/98	1924/25		
Clark	—	—	1850	1875	1900	1929		
Dupriez	1789/92	1808/14	1846/51	1872/73	1895/96	1920	1939/46	1974
Rostow	1790	1815	1848	1873	1896	1920	1935	1951
Mandel	—	1826	1847	1873	1893	1913	1939/48	1967
Van Duijn	—	—	1845	1872	1892	1929	1948	1973

Source: After Van Duijn, 1983

impossible; without the profusion of output in basic consumer goods, such as cars and appliances, it would have been unnecessary (Aglietta, 1976; Dumenil et al., 1987). We also depart from the Regulationist emphasis on internal causes for failure of the Fordist regime, such as an inability to sustain high rates of productivity increase and the unraveling of the mass-consumption wage bargain; instead we emphasize the continued dynamism of capitalist industrialization as the chief cause of destabilization. That is, continued investment in the general expansion of production intensified international competition and led to vast overcapacity in many sectors; new products undermined markets for many older industries; and new production methods, such as those of Japanese car manufacturers, eroded the competitive position of once-dominant firms in North America and Western Europe.

By the same token, we cannot agree with the stress which the *Monthly Review* school puts on chronic stagnation due to the excess capacity and monopoly prices of large corporations (Baran and Sweezy, 1966); nor is the URPE (Union of Radical Political Economists) school convincing in arguing that labor militancy and bureaucratic sclerosis within large corporations brought about the demise of post-war US prosperity (Bowles et al., 1983). The monopolistic position of US capital in the mid-twentieth century depended heavily on the advantage that US productive capacity lent to American companies; it has faded appreciably along with the diminishing productivity gap over the years, and collapsed completely in overbuilt and technologically stagnant US industries such as steel. Similarly, high wages, rigid work rules and management featherbedding did not precipitate a crisis for US, Canadian or European capital; they only became significant costs, and a target for capitalist attack, once markets and profits had been seriously eroded. As we argued in chapter 2, wage rates and other input costs influence profit rates, but they are secondary to the disequilibrating effects of investment, competition and technological change in explaining the dynamics of capitalist growth and crisis.

The timing, as well as the cause, of the epochs of capitalist growth associated with regimes of accumulation remains controversial. These periods are generally thought to correspond to long waves in economic activity, measured by output expansion rates, employment growth and the like. Various datings of long waves of growth are shown, for comparison, in table 7.1. It is unlikely that such dates would be perfectly synchronous among capitalist countries. For the United States, a reasonable periodization would run circa 1790–1845, 1845–1900, 1900–40 and 1940–90 (Walker, 1977).

7.3.3 The geographical imprint of changing production ensembles

Each ensemble of productive forces brings with it a characteristic form of spatial organization. First, each period has its key loci of growth, or localization patterns, as a result of the new alignment of factor inputs, markets, and productive know-how. Each ensemble of dominant production sectors comes to be centered on particular global regions: textiles in the North of England; cars and consumer durables in the US Midwest; and electronics in the US Sunbelt and Japan. Second, new production ensembles will disperse over time to subjacent zones and secondary national cores, as the leading industries grow and innovations spread, as with the appearance of important textile centers in the mid-nineteenth century in northwest and southeast France, Catalonia, New England, and Philadelphia. Third, the geographical core of older industries may shift as the new technologies transform old products and production methods, as when the radio became transistorized and jumped to Japan or when shipbuilding shifted to supertankers built in Korea and Japan. Finally, areas closely associated with older technologies will stumble, even collapse on occasion, if their key sectors do not relate closely to the new technologies – thus the disappearance of the Severn Valley centers with the passing of the iron age or the gutting of Manchester and environs as the textile mills finally expired *en masse* in the North of England.

A sketch of five long waves of territorial industrialization in the United States, corresponding to the technological ensembles previously enumerated, might look like this:

- The long wave of the first Industrial Revolution, 1790–1845: workshop industries arose in mercantile centers of the eastern seaboard. In New England, textile factories ringed Boston and Providence; gun, clock and hat making were found in Connecticut; shipbuilding and fishing towns dotted the coast; and the shoe industry clustered around Boston, from Worcester to Lynn. Iron-making was scattered throughout Pennsylvania, and Philadelphia was the early national leader in metal-working. Cincinnati became the queen city of the old Northwest Territories and center of meatpacking, while New Orleans mushroomed on the strength of the Mississippi River trade by flatboat and steamboat. New York City took the lead away from Philadelphia in trans-Atlantic trade and financial transactions.
- The long wave of steel and machine-building, 1845–1900: New England continued its industrial dominance by perfecting the "American system" of machining, leading in such industries as bicycles and clocks; Philadelphia also developed vigorous textile, chemical and machining complexes, with New York not far behind. New York City riveted its hold on garment-making, publishing, transoceanic shipping and finance. Pittsburgh, in the bituminous coal zone, became the steel and glass center and vital entry point to the Ohio Valley, while Buffalo became a strategic industrial center at the junction of the Great Lakes with the Mohawk valley corridor. The upper Midwest began to industrialize chiefly through

resource-processing industries such as grain-milling in Minneapolis, copper smelting in southern Michigan, brewing in Milwaukee, and oil refining in Cleveland. Chicago burst onto the world stage engorged by slaughtering, machine-making, and railroading. Northern California made its appearance as a mining complex, wheat exporter, and mercantile center for the Far West.

- The long wave of early Fordism, 1900–40: the Midwest became the country's industrial core led by the automobile ensemble centered on Detroit and the machine-tool complex focussed on Chicago. These were joined by integrated steel mills along the Great Lakes, and household appliance assembly all through the region. The radio, movie, phonograph, recording, and broadcasting industries grew up in and around New York City, while Niagara, New York and Philadelphia were centers of the chemical industries, including pharmaceuticals and cosmetics. Electrical equipment concentrated around Boston, the Hudson Valley and Pittsburgh, business machines spread into upstate New York, and telephones were centered in New Jersey. Meanwhile, the oil, movie and aircraft industries became heavily concentrated in Southern California after promising starts in locales in the Northeast. Irrigated agriculture, canning and large-scale construction also developed into California specialties. The South took over as the textile center of the country, lumbering and wood products moved to the Pacific Northwest, and non-ferrous metal mining and smelting shifted to the inter-mountain West.
- The postwar wave of high Fordism, 1940–75: the Midwest continued to dominate as US mass-production industries commanded world markets across an astonishing range of sectors; yet, even as the system reached its apex, US multinationals increasingly located new plants in Europe, the southeastern states, and the Third World. Oil drilling, petroleum refining, and petrochemicals came to be centered on Texas and Louisiana, especially around Houston. Aluminum and aircraft became mainstays of the Pacific Northwest, while timber was exhausted in a final binge of cutting, and began to shift back to the South. Irrigated agriculture spread vigorously, but especially to Arizona and Florida. Meanwhile, Southern California gained three more propulsive industrial groups: missiles, aerospace electronics, and television broadcasting. The San Francisco area not only gained electronics, but jumped ahead of Chicago in finance, as well as becoming a national leader in tourism. Electronics has also grown up around Boston, the Hudson Valley, Minneapolis, Dallas and Phoenix.
- Post-Fordist Industrialization, 1975–: Indications of a new era of territorial industrialization are already before us, so it is possible to speculate on the new configurations. As the Fordist industries have suffered overproduction and lost out to stiffer

worldwide competition, the Midwest has undergone severe de-industrialization. Armed with new production methods, the Japanese have shifted the world centers of many industries to Asia. The US industrial core has begun to reshape itself into three distinct, noncontiguous areas: the Southwest, with California at its center; a revivified New England; and the central business districts of a few very large metropolitan areas. The dominant post-Fordist production ensembles (high-technology industry, producer and financial services, and revitalized craft industries), using flexible production methods, are coming to be centered in these areas. Still newer sectors appear to be emerging, based on supermicroelectronics, laser optics, bioengineering, and the like, but their locational patterns are as yet unclear. Finally, many older sectors are being transformed by the application of flexible production methods and Japanese management techniques, and consequently are undergoing spatial reorganization.

7.4 Conclusion: geographical industrialization and territorial growth

Frederick Jackson Turner's famous thesis about frontier development in the United States, while burdened with a good deal of romantic baggage, evokes something essential in the American experience of renewal through territorial invasion and conquest. What Turner (1920) did not see was that the heart of the matter lay not in the land nor in the "soul of a nation," but in capitalist industrialization. He also did not anticipate the way that new frontiers might yet be opened within the world economy.[11] A number of modern writers have suggested that territorial expansion is a principal source for the renewal of capitalist accumulation, in economic as well as political terms (Norton, 1986). Di Tella sees the frontier as a disequilibrium state "bursting with business opportunities with big profits and economic excitement" (1982, p. 15); in this account, the scramble for superprofits in a growing region generates a powerful stimulus to investment and to innovative behavior. Harvey (1982) portrays capitalism as a system prone to crisis, in which "the spatial fix" offers a way out of the dilemmas of capital imprisoned in the spatial configurations of the past. Smith (1984) analyzes the way that "uneven development" offers critical geographical differences to be exploited for new sources of rapid capital accumulation. While this is a beginning, it does not go far enough in filling out the solid forms of production that make up the process of territorial industrialization.

[11] An early argument about the closing of the capitalist world economy was fought out about the same time between Luxemburg (1913) and Lenin (1917).

As we have shown, geographical industrialization generates the large patterns of uneven territorial development. First, spatial expansion results from the generation of new industry growth centers. Second, differentiation of places is tied to the spatially selective channeling of growth by particular industries. Third, instability in territorial fortunes results from a combination of extension into new places, abandonment of older territories, and re-differentiation of industrial locales. All these effects are amplified when seen in terms of waves of industrialization based on new production ensembles. Thus, territorial extension, differentiation, and instability are not afflictions visited on industrialization but conditions upon which capitalist development thrives.

8 Economy, Society, Territory

8.0 Introduction

So far we have focused almost exclusively on production and its development through the process of industrialization. We turn now to the ways in which the territorial implantation of industry is wrapped up in the distribution of income and power between classes. First, we present a case for the spatial contingency of class income, by demonstrating that the division between profits and wages is determined in light of the territorial division and integration of production. The same holds true for class expenditures, or the division between investment and consumption. Further, we distinguish among territorially specific regimes of accumulation that combine particular relations of production and distribution. To breathe life into this framework, we next introduce the active conflict, maneuver, and reconstitution of classes in and through the process of territorial industrialization. As new propulsive sectors periodically renew the productive base of capitalism, they also offer the possibility of new relations of distribution and political power. Experiments in establishing new regimes of accumulation, therefore, take place through territorial change, and as these galvanize business energies and spread, they reinvigorate the capitalist system as a whole. In short, the historical development of capitalism cannot be understood without considering the ways in which the forces and relations of production are shaped in differentiated territorial contexts and geographically diffused throughout the world.

8.1 The macroeconomics of the territorial division of labor

Geography is central to labor supply differentiation and to the existence of segmented labor markets. Laborforces are differentiated along several dimensions which may interact in complex ways: skills, work discipline, standards of living, work life prospects, bargaining power. This differentiation of employment relations makes possible a range of locational choices for economic activities and heightens the degree of territorial social differentiation generally. We have already considered the effects of this heterogeneity on the location patterns of individual industries. In this section we take up the collective effects of the territorial division of labor on distribution and accumulation.

8.1.1 Wages, profits, and location

Let us consider further the relation of location to wages and profits, or class incomes. Profit rates – whether at the firm, sector, or economy level – are inversely related to the level of wages, all other things being equal (Pasinetti, 1977). The differentiation of employment relations, and thus of wages, describes a field of possibilities for profit making. The regional mosaic of wages (for a given performance level) thus establishes the *ex ante* profit frontier for an industry (Sheppard and Barnes, 1986). It follows that different locational configurations would maximize profits for different patterns of regional wage differentials. In seeking locations for production and in forming employment relations, capitalists establish actual distributional relationships associated with the laborforces available at different sites. In other words, location is a means to adjust the wage by exploiting territorial variation in labor supplies (Warntz, 1965). With inter-regional wage differences not precisely corresponding to inter-regional labor skill or intensity differences, location becomes a point of leverage in the capitalist effort to increase profits.[1]

This is not to say that capitalists' locational strategies are likely to be optimal. Individual profit-maximizing locational decisions do not necessarily maximize aggregate profit rates. *Ex post* profit rates depend on the interaction of firms and sectors, including locational interactions. As a result, capitalists in a given industry may, individually or collectively, fail to maximize profits for a given wage level or fail to produce a locational pattern convergent with the optimal wage-profit frontier. Indeed, game theory and the logic of collective action suggest that such ideal results are remote: even if an optimal geography of production were initially established, there would be nothing to prevent individual producers from relocating so as to lower their own wage bills, undercut competitors, and reap superprofits in the short run. Conversely, there is nothing to assure that movement from a suboptimal to an optimal pattern will occur (Barnes

[1] This reasoning would hold even if wages varied simply in direct proportion to different skill levels. In that case, locations would be selected according to skill requirements, with only the absolute levels of profits in the economy as a whole varying according to the center of gravity of the aggregate wage structure. Profit rate equalization requires an additional restrictive assumption: skill-productivity levels would have to be fixed along some common scale between industries and regions. Such an assumption is inconsistent with our conception of capital. Note, too, that an ideal capitalist would not merely find the optimal (i.e., profit maximizing) location pattern for a predetermined wage level, but would find the location pattern which represented the optimal wage-profit frontier (Sheppard and Barnes 1986).

and Sheppard, 1985). The social average profit rate is a locationally-contingent fact, but not necessarily a maximum for capitalists nor an equilibrium. The same is true for the collective wage bill: even though location is a weapon in the capitalists' arsenal, their locational behavior cannot insure that this bill is held to the minimum. Nonetheless, the basic point holds: location is a principal strategy in the effort (however misbegotten) to maximize profits and minimize wages for a given level of worker performance.

Class conflict over the rate of exploitation will be fought out within the context of differential potential profit rates owing to technological and locational divergence among industries. High rates of profit and growth, for example, generally create high ability-to-pay; but they will meet a wide variety of locational and, hence, labor supply possibilities from sector to sector. If a potentially high-wage industry is unable to relocate and so weaken its workers' bargaining strength, it may find itself forced to concede much of its potential profit to real wage increases. San Francisco's construction workers, for instance, secured some of the nation's highest wages around the turn of the century because they found themselves in a relatively autarchic labor market and were able to secure an unprecedented degree of workplace and political power (Kazin, 1987). On the other hand, a key element in California agriculture's high profitability has been the success of employers in recruiting waves of cheap and powerless immigrant labor (McWilliams, 1949).

Industries in which profit rates are low or declining owing to long-run technological forces can also use locational maneuvers to continue to exploit gaps between laborforce performance and wages in different regions, but not all are free to do so. Some try to revive flagging profits by replacing workers with new immigrants, moving beyond the reach of unions, or seeking areas with low-wage traditions (Bluestone and Harrison, 1982); they may perform very well for a time on this basis, as exemplified by the current expansion of garment production in London or Los Angeles. Others find themselves unable to escape or defeat militant laborforces, and are forced to cut back production and lay off workers, as in the French auto and steel sectors over the last decade (Preteceille, 1985).

In short, locational behavior is likely to make profit rates more differentiated than they otherwise would be (cf. Vietorisz and Harrison, 1973), i.e., sectoral average profit rates are contingent upon location. We can take this one step further. The performance of all sectors in the economy establishes the average profit rate of capital as a whole and this depends, in turn, on profit rates established, sector by sector, through choices of location and laborforce. The social average profit rate is a geographically-contingent fact. We shall return to the distributional effects of locational patterns, but for the time being it is enough to establish that locational maneuvers and place-based class struggle in the many sectors of industry can affect wage and profit rates.

8.1.2 Investment, consumption, and location

Relations of distribution do not stop once a wage rate is established at the point of production. The division between profits and wages determines the funds available for investment and consumption, or class expenditures. The level of class expenditure can affect the fortunes of any industry. In a simple two-class, two-sector model, lack of sufficient consumption (or investment) will bring a failure to sell all the wage (or capital) goods produced and actual revenues will not cover promised incomes; this is the problem of "effective demand," in Keynesian terminology, or "realization," in Marxian language. In a dynamic setting, a high rate of investment will also increase the rate of productivity growth, in most circumstances.

Crisis-free capital accumulation – the Marxian term – or balanced economic growth in a capitalist system – the Keynesian phrase – depends on the double relation of profits/wages and investment/consumption.[2] For growth to continue, two requirements must be satisfied: first, workers' income must grow at the same rate as the output of wage-goods industries increases; second, capitalists' income and investment must grow at a rate sufficient to cover the increase in output of capital goods. But beyond this, the rate of investment must stimulate just the right amount of productivity growth in both sectors to match investment and consumption expenditures in future periods.[3] Moreover, the long-run growth performance of the economy rests both on the ability to avoid crisis (which squanders resources and time) and on the underlying trajectory of growth (which may be set at a low-profit, low-investment rate or at a high rate, depending on a variety of conditions).

Economic growth can be achieved at any one of several possible combinations of wage/profit and consumption/investment rates. For simplicity, we can distinguish four common national models of capitalist growth in recent times. The first, which may be called the Japanese model, combines modest wage and consumption levels, high profit rates, and very rapid accumulation; exports make up for the relatively low per capita internal consumer demand, and growth rates are elevated by productivity-expanding investments. The second, or Brazilian model, consists of very

[2] Put in Sraffan (or neo-Ricardian) quantity terms the matter reads thus: "Essentially, all producers must increase their output at the same rate per production period; and the relative quantities of each good produced should be such that this period's supply exactly provides for the next period's demand . . . [T]he significance [of this] is that there is a unique set of relative production levels that solve the equation, and permit an equal level of growth for each sector in each region. For the wage-profit frontier upon which the location pattern is based, a consumption-growth frontier can be derived, showing every possible division of the economic surplus, defined in physical terms, between consumption and investment" (Barnes and Sheppard, 1985, p. 13).

[3] If we assume perfect price adjustment in neoclassical fashion, the problem disappears – but this is an unreasonable assumption because prices cannot adjust quickly and accurately enough to match the shifting contours of the economy.

low wages, relatively high investment, mass consumption by the upper classes (a combination of capitalists, salaried middle class, and rentiers from the landed class), and erratic macroeconomic performance with a slow secular expansion tendency. The third, or North American and Northern European model, combines high wages and elevated worker consumption with moderate profits and moderate rates of investment, such as prevailed over a period of very strong post-war growth.[4] A fourth, or British model, juxtaposes modest wages, moderate mass worker consumption, poor profits, weak investment and slow growth (Anderson, 1987). There is, in short, no single path to growth, and nothing to insure that the proportionalities necessary for balanced macroeconomic growth will be achieved. Similarly, there is no assurance that balanced growth is rapid growth.

The geography of employment relations bears on the investment-consumption relationship as well. The territorially-contingent struggle over the social average rate of profit, or the distribution between capital and labor in production, puts certain bounds on national or regional investment and consumption. In contemporary Brazil, for example, industry is extremely concentrated in and around São Paulo, and wages in that agglomeration are subject to continuing downward pressure by the reproduction of a labor surplus due to immigration – itself chiefly a consequence of aggressive modernization of the agricultural countryside. When labor militancy threatens, selective deagglomeration to even lower-wage and poorer regions sets up a demonstration effect within and between industries. These wage dynamics, in turn, severely restrict the expansion of the society's consumption patterns and reproduce structural bottlenecks in the Brazilian industrial system which retard self-sustained Fordist growth of the sort realized in post-war North America and Europe. Thus, even though Brazil's industrial economy has well-developed forces of production, geographical differentiation underpins social relations that restrict Brazil to a more "underdeveloped" macroeconomic growth path (Storper, 1984).

Quite the opposite occurred in the United States as it entered into its Fordist period. In the 1930s, successful unionization drives in the great industrial cores broke the wage-reducing, anti-union strategy of US business (the "American Plan") that had helped plunge the country into an overproduction/underconsumption crisis (Devine, 1980). With increasing solidarity in both the workplace and the community, the US working class was able to pave the way for postwar collective bargaining

[4] This has given way to a pernicious degree of falling wages and worker consumption, rising capitalist and upper-middle class luxury consumption, high profits but low investment as surplus is squandered on consumption, mergers and financial speculation (Davis, 1986).

agreements that linked rising wages to increasing productivity in a virtuous spiral that lasted for almost thirty years (Davis, 1986). In the same period, massive suburbanization of production and jobs was accompanied by an equally dramatic residential suburbanization, and the consumption bundle associated with this move was the major stimulus to demand for the outputs of the mass production industries which dominated the American economy; the "suburban solution" was a geographically-based means of coping with the problem of effective demand and labor militancy under Fordism (Walker, 1977; Gordon, 1977).

8.1.3 Distributional relations and regimes of accumulation

We have previously introduced the concept of a regime of accumulation to capture the set of political-economic relations that arise with and sustain major epochs of capitalist growth. Income distribution and class expenditure are central to the ensemble of growth-sustaining conditions of every such regime. Now it can be seen that regimes of accumulation are established only as the result of territorially-differentiated class practices and distributional outcomes. The territory is essential to the marshalling of social, political, and institutional resources needed to provide an environment tailored to the growth and development of industrial ensembles. Such place-specific regimes of accumulation may be said to represent particular 'territorial capitalisms' (Morishima, 1982; Dore, 1987). The geographical particularity of regimes of accumulation is due in large part to the large-scale social, institutional and political arrangements of places—hence the range of regimes built on much the same technological foundations in different areas of the capitalist world.

Regimes of accumulation can exist at virtually any territorial level: regional, national, or global. The nation-state is the principal unit for a workable set of production and distribution relations and for constructing regimes of accumulation (Lipietz, 1986); it is the gravitational force that most tightly binds the class practices and institutional arrangements of politics, economics, and territory. At a minimum, the national state remains the boundary of completely free trade, the field of operation of currencies (and thus discrete price systems), and the territory of relatively homogeneous cultural and linguistic communities. Because no two countries regulate social action in identical ways, important differences remain, despite the apparent unity of even the postwar Fordist regime. At the global level a hegemonic role is occasionally played by leading national economies and states in regulating world trade, projecting their currency as the international standard, and patrolling world commerce with their gunboats. This was the function of Britain in the nineteenth and early twentieth century, and of the United States after World War II

(Block, 1977).[5] Finally, even subnational regions can have a significant place in the formation of distinctive territorial capitalisms. Subnational differences in social regulation are especially marked in countries with a relatively high degree of decentralization, as in Canada and the United States. For example, California is a distinct region, not only in its unique mix of industries but also in the peculiarities of its labor struggles, political makeup, commitment to science, consumerism, and a boosterist, get-rich-quick, cure-all culture (McWilliams, 1949; Davis, 1987).

8.2 Territorial industrialization and class relations

We now move beyond income distribution and economics, narrowly construed, to the broader relation between territorial industrialization and class. We shall argue that geographical differentiation is central to the concrete experience and formation of class and to the balance of class power. Class relations are repeatedly revised in light of the shifting geography of industrial development.

8.2.1 Territorial differentiation and the balance of class forces

In the process of forming differentiated territories, industries also lay the foundations for opposing processes of fragmentation and class unity, and thus alter the balance of class power in capitalist society. That is, we address ourselves to the place of location in what Gramsci (1934) called "the war of position" between labor and capital. As this is most readily apparent with regard to the working class, we shall begin with a simple model of the relative strength of labor in wage bargaining in geographically varied markets.

The ensemble of territorial employment relations and the economy-wide system of labor market segmentation engenders in the laborforce a sense of difference and stratification that is difficult to overcome. Labor market segmentation has generally pernicious effects on the wages of the working class as a whole, not just those at the bottom of the heap. Segmentation tends to pull down the relative benefits of the most favored groups of primary workers toward those of the least favored, rather than scattering wages randomly around an average (Reich, 1981). Territorial difference and dispersion works against class solidarity in several ways: communication is more difficult, different local unions develop, cultural

[5] Without a hegemon, world capitalism tosses on rough international seas, as happened in the period between the wars as Britain faded and the United States refused to take up the mantle of power; this is apparently occurring again today with the passing of economic initiative from the United States to Japan, while political power remains disproportionately in US hands.

separation becomes more marked; finally, workers may come to identify with place over occupation, industry, or class. Such workers can more easily be set up by employers to compete with similarly-qualified workers in different local labor markets (Mann, 1988). As a result, territorial labor market segmentation appears to bias class distribution toward capital and away from labor, all other things being equal. The social average profit rate can thus be said to be geographically-contingent beyond the effects of the locational behaviors of individual industries. The spatial pattern of industry distribution and territorial labor markets affects the distribution of income between classes by altering the relative bargaining strength of labor and capital as a whole.

The working class is not without means to overcome space and internal disarticulation, however. Wage rates in local labor markets are sensitive to wage rate changes among firms and sectors in the locality, beyond those resulting from changes in the regional cost of living. These are the wage pattern effects referred to in chapter 6. Workers tend to engage in pattern bargaining, taking a lead from other workers in forming wage expectations and making contract demands. Pattern effects are most clearly in evidence in industries consisting of large multilocational firms; conversely, they are weakest in industries where neither production systems nor product markets extend across regions, such as construction. Pattern effects are found in occupational groups as well, particularly those requiring similar skill levels, and most clearly in professions requiring formal credentials.

We can suggest two types of territorial pattern effects. First, workers within the same industry, even in different occupational groups, will coordinate their wage demands more if that industry is territorially concentrated than if it is territorially divided. Second, workers in different industries may follow each other more if they are territorially concentrated than if they are divided among many different labor markets (Hamermesh, 1975). Territorial pattern effects raise labor's share of sectoral and national income. They tend to reduce differences between high- and low-wage industries and between high- and low-wage occupations within industries (cf. Roemer, 1979; Reich, 1981). There are strong counter forces, however, including sub-local labor markets and the changing industrial structure of the local economy. Territorial patterning seems to be most marked with a relatively stable local economy that allows workers to build experience, political subjectivity, and institutions to deal with employers and the state; it appears weakest in times of rapid growth and in-migration (cf. Barsly and Personich, 1981; Freeman and Medoff, 1984). There is even stronger evidence that in countries where industrial unions reach across regions, wage rates are nationally patterned, and the state enforces labor bargaining (as in European social democracies during postwar expansion), the working class as a whole gained. Many European factory regimes became national in scope because once rules were worked out in some places, there was strong pressure to impose those rules nationally to foreclose the possibility that more militant regional workforces or more conservative regional employer groups would refuse to go along.

National factory regimes in Europe, then, did not spring directly from nation-wide conflicts; instead their formation was rooted in the search for systems that would rationalize and regularize the particular dilemmas posed by regionally-specific and sectorally-specific production politics.

In the United States, on the other hand, no firmly affixed national factory regime has ever existed, even at the height of postwar prosperity. The Wagner Act of 1935 established the union local as the bargaining unit, and national-level contracts must be achieved by organizing all units of a firm and seeking ratification at the local level. The Taft-Hartley Act of 1947 allowed states to develop their own local labor legislation; twenty states, mostly in the South and West, have passed so-called "Right-to-Work" (open shop) laws since then. The Davis-Bacon Act of 1936 specifically requires that local labor market conditions prevail in federal construction activity. These and other protections of local variations present an obstacle to achieving uniform national standards and bar certain kinds of territorial pattern effects (Clark, 1983, 1986).[6]

Geographical maneuvering is strongly in evidence in the recent redistribution of income between labor and capital in the United States. In the 1970s, income distribution began to become more unequal for the first time since World War II, in what Harrison and Bluestone (1988) call "the great U-turn." Inter-state income levels have also taken a sharp turn toward greater inequality (Amos, 1985). Relocation of employment appears to have been a major source of this redistribution (Clark et al., 1986; Clark, 1989). First, decentralization of branch plants in older industries in the 1960s and 1970s undermined union strength in the Midwest and contributed to the demise of the Fordist regime of accumulation (Schoenberger, 1987b). Then, the emergent industrial areas of the 1980s established a very different wage regime of bifurcated rewards with a "missing middle" – a split made possible chiefly by the absence of unions among lower-tier workers. At these frontiers of class redistribution, cheaper labor is not simply sought out, it is actively created through militant anti-unionism. The growing electronics clusters are notable for a dualized wage structure, a union-free environment, feminization of the manual workforce, and hiring through temporary labor or worker-leasing agencies (Pfeffer and Baron, 1986; Scott, 1988b). Office industries are also characterized by dualized wages, with millions of low-paid and part-time jobs for female clerical and minority service workers, while (largely) male managers, brokers, lawyers and the like enjoy incomes that have spawned the gentrification of many central cities (McMahon and Tschetter, 1986).

[6] Unfortunately, there is little aggregate evidence about the ways territorial pattern effects work (Warntz, 1965). In particular, we need to know much more about the specific effects on wage dispersion of sectorial combination and division in space, and how interregional wage differences develop as a consequence of sectoral composition and the political experiences built up in local labor markets with different combinations of sectors.

8.2.2 The social frontiers of growth

Not only do locational windows offer capitalists new possibilities for unleashing the growth potential of specific industries, they provide the opportunity to establish class relations that improve capitalist fortunes, galvanize entrepreneurial energies, and allow for the shaping of economics, politics, and culture in ways that reinvigorate the conditions of accumulation (Di Tella, 1982). New and restructured industries are frequently accompanied by new groups of capitalists, new labor relations, new forms of work organization, new groups of workers, and new patterns of consumption. Capitalism as a whole grows and changes through such regionalized growth frontiers of experiment and change.

A growing literature on innovation and regionalism suggests that entrepreneurial capitalists, consciously or unconsciously, generally seek to avoid perceived social and institutional rigidities in well-developed regions (Scott, 1985, 1988b; Hall and Markusen, 1985; Clark, 1986). These rigidities include conditions that might be identified in conventional models, such as "factor oligopsony" when one or two industries dominate a regional economy (Chinitz, 1961), but they have more to do with social practices and class relations than with market transactions. At the simplest level, an entrenched business class may present a social barrier to the establishment of a new group of entrepreneurs associated with new product groups (Schumpeter, 1942). Existing social networks monopolize the attention of creative individuals, funding sources, and local government (Markusen, 1985). Managers and investors become deeply bound to established ways of doing things and are blind to new possibilities (Morgan and Sayer, 1988). In France, for instance, it was an article of faith among bankers for many years that the computer industry was not going anywhere, and therefore budding entrepreneurs tended to start up software firms because they required less capital than hardware production. The President of General Motors, asked by students at Harvard Business School about possible inbreeding in GM, replied that upper management was a highly diverse group drawn from universities all over the Midwest![7]

Entrepreneurial capitalists, then, often gravitate toward regions where the slate is relatively clean, and situate themselves at the center of a growing economy and community, so they can participate in shaping the area's social, institutional, legal, and political traditions. The US business class is usually deeply involved in political growth coalitions at the regional level (Logan and Molotch, 1986). These coalitions typically take several lines of action in trying to mold the regional political economy. One is to commandeer local government by participating in electoral campaigns, or on the commissions, boards, and other non-elected regulatory bodies that exert considerable influence over decisions important to regional

[7] Thanks to Martin Manley for recounting this episode.

development (Mollenkopf, 1983; Hartman, 1984). Another strategy is to form extra-governmental bodies – such as the "opinion leaders" clubs found in most cities – that function as prestigious centers for the manufacture of public opinion and work to develop consensus among those with power; examples include the Commonwealth Club in San Francisco and the Commercial Club of Chicago (Domhoff, 1979; Heiman 1988). Newspapers and the broadcast media are usually central players in local growth consensus. Private foundations are also very important in funding philanthropic activities and the institutions that influence public perceptions, and in funding the research which defines legitimate discourse about the nature of regional problems and processes. To speak of the class character of growth, then, is not to refer simply to a "favorable business climate;" it includes the political and institutional means by which growth is organized and ideologically represented (Friedland and Bielby, 1981).

Labor and employment relations pose perhaps the greatest obstacles to capitalists' freedom to operate at will. We have noted that capitalists at times try to use internal and external solutions to establish a clean slate with respect to labor relations, work rules, work traditions, and institutional and political relations between capital and labor. Taken as a whole, the emerging post-Fordist epoch is characterized by lower rates of unionization, greater segments of non-union employment, more negotiable work rules, reduction of internal labor markets to a smaller core workforce, and lower rates of militancy in general. This re-formation of employment relations would have been unimaginable without the massive inter-regional and international movement of industry in the 1970s and 1980s. Workers in already-developed regions would very likely have tried to bar the necessary technological, organizational, and employment innovations. Mighty IBM offers a valuable example of long-term growth through a combined external and internal solution of extreme labor control and flexibility. Despite superficial parallels with General Motors and other large companies of the Fordist era, IBM has prospered with the help of a militantly anti-union corporate culture that it has implanted in virtually self-contained growth peripheries. These are discreetly positioned away from existing industrial centers, as at Oak Ridge, Tennessee or Boca Raton, Florida, or as at its European plants near Montpellier, and outside of Rome, Milan, and Glasgow; even in non-union Silicon Valley, IBM's huge plant is twenty miles south of the industrial core – and a new one is being added 20 miles further south still – to increase the social distance from the San Francisco Bay Area's labor and political climate.

Conversely, in places where capital finds itself strapped by a laborforce with strong traditions of workplace performance and solidarity, productivity is likely to change at a much lower rate, and capital may evade an entire country by transferring investment abroad. An examination of hourly worker productivity in the advanced capitalist countries in the postwar period, for example, shows Britain lagging in almost every sector

(Williams et al., 1983). It is fair to speculate that the long decline of British manufacturing competitiveness is in part due to the power that British unions wield over the labor process, a power strongly rooted in rules established very early on in the Industrial Revolution. British labor process norms are not easily accommodated to contemporary productivity requirements, and are defended fiercely and ably through shop floor organization and class militancy (Anderson, 1987).[8]

The story of reconstituted regimes of accumulation is not a matter of extensive geographical frontiers alone. Any place, any difference may do. The territorial differentiation and bounding of social life are among the principal determinants of the strategies employed to form and reproduce social life (Giddens, 1984). The partitioning of social interaction in space as well as time creates a variegated grid of uneven development that continually offers capital new frontiers to enhance its rate of profit and its class power (Smith, 1984). The constant flux of economic activity in search of new differentials to exploit has an important dynamic effect on capitalism as a whole.

8.2.3 The diffusion of new practices

New territorial capitalisms are vital to the periodic renewal of industrialization and economic growth, as in the case of Japanese capitalism or American capitalism before it. We refer here to the competition among places, which is every bit as important to capitalism as the competition among firms (Harvey, 1985; Logan and Molotch, 1986). The success of areas with new regimes of accumulation puts pressure on older territories to reshape their social structures to compete with new growth centers for investment and market shares. As lagging industries and their regions come to emulate new regimes of growth, reverse diffusion of social and political conditions of growth takes place.[9] What begins as a specific politics of place comes to be a politics of wide areas of a nation or of the capitalist world in general. We can point to five facets of this diffusion of class practices.

First, new forms of management and work organization tend to be developed in association with new products, sectors, and technologies. These provide profitable examples to capitalists in older regions and to other countries, as witness efforts to emulate the Japanese model in factories throughout North America and Europe (Morgan and Sayer, 1988). Even a nominally Socialist government in France in the early 1980s

[8] Poor management is also to blame for Britain's performance, of course (Morgan and Sayer, 1988).

[9] Diffusion can occur through any of the spatial growth and integration processes previously identified. For example, industries attracted to new markets may find the new region's social structure unusually hospitable and be inspired to emulate these new-found productivity arrangements in other locations. Similarly, subcontractors or assembly units of an industry may develop successful labor practices that draw industries to a formerly peripheral territory, as has happened along the United States–Mexico border.

advocated a strategy whereby the French should: "beat the Japanese on their own ground and with their own methods. Adapt to France the social innovations that underlie the performance of a country such as Japan. This is the path to follow."[10]

Second, new models of income distribution are established. California, for example, has a strongly post-Fordist class structure, with a greater proportion of income going to the upper third of the population than in any other region of the United States (Davis, 1986). This provides an example for income distribution that contrasts sharply with the older, midwestern model of a solid middle filled by a unionized working class. The Atlantic Frostbelt areas are now beginning to see something similar to California, in the new American "bicoastal economy." Such revisions of income distribution are not simply the results of changes in the industrial mix and the occupational structure, but are predicated on the ability of new industries to develop revised wage structures in regions where they are acceptable, and thus to discipline older, more resistant areas by the success of their growth and high investment rates.

Third, inter-regional diffusion can reconstitute class politics. In the United States, for example, the national institutional gains of the labor movement ceased in the 1950s, largely because southern right-to-work states, granted the power to deviate from the national pattern, provided an opening for a tidal wave of investment from non-unionized companies. Furthermore, the southern congressional delegates have been able to block reform of national labor law by virtue of the South's growing population and economic importance (Bensel, 1983). In addition, there has been considerable pressure on heavily unionized states to take local actions to weaken any institutional protections afforded labor; even in unionized industries and union areas, workers are increasingly reluctant to initiate representation elections or grievance procedures, and rates of failure in union elections are converging toward those in the right-to-work states (Clark and Johnston, 1987; Clark, 1989).

Fourth, a successful regional model may have great impact on production politics and shop floor practices. Much of the debate over employment in the 1980s has been about work rules and their effects on worker performance and labor allocation in production (OECD, 1986). Nonetheless, the limitations on employer flexibility imposed by union-won work rules did not bother employers as much before they felt severe competitive pressure from abroad. Nowhere is this clearer than in the mimicry of Japanese methods of personal address and occupational classifications in such experiments as the GM-Toyota joint venture in Fremont, California, or GM's announced intention to use Japanese work rules in its Saturn Division complex in Spring Hill, Tennessee, and its assembly plant in Van Nuys, California.

[10] Government official quoted in *Libération*, August 18, 1985, p. 10 (authors' translation).

Fifth, forms of state intervention have diffused from one country to another. In the 1950s, virtually all the advanced capitalist nations of Western Europe and North America adopted some form of Keynesian macroeconomic regulation and welfare-state policies. In recent years, most of these same nations have backed away in substantial measure from these ideas. The current vogue for neoconservatism and fiscal retrenchment did not arise spontaneously or simultaneously at all places, however; its origins lie in specific places. Japan and the newly industrializing countries are conspicuous for combining low social wages with high productivity. Their relative success has imposed a discipline on welfare-statist governments. Likewise, the ideological and organizational origins of neoconservatism – idealizing entrepreneurism, privatization, and minimal government – did not arise ubiquitously; instead, their most important collective and organized expression began in particular places such as Southern California, home of Ronald Reagan, and the City of London, bastion of Thatcherism.

To emphasize the local origins of state intervention is not, however, to claim that states are independent of the forces of a global capitalist economy. Indeed, quite the opposite conclusion may be drawn. Keynesian welfare statism was possible in part because of the enormous surpluses being generated in the post-war period. With the breakdown of Fordism, competition has sharply reduced the power of any individual capitalist country – whatever its size or wealth – to carry out a welfare-state economic and social policy. The failure of the French Socialist government in the early 1980s is a striking case in point (Singer, 1987). It is thus consistent to hold that models of state intervention stem from particular localities, but that competitively efficacious models impose themselves on other localities, sharply circumscribing (though hardly eliminating) the power of those localities to deviate substantially from the competitive requirements of accumulation.

8.2.4 The geographical formation of classes

The territorial differentiation and renewal of capitalism do more than affect relative incomes and relative power in the class struggle: they affect the constitution of classes themselves. This is fundamental to the new territorial capitalisms, and regimes of accumulation, that periodically appear and shake the foundations of growth. Just as industry has been more or less continually transformed, so, too, have the bases of class formation. This is not to draw an exact parallel: Marx was at pains to contrast the relative stability of the social relations of capitalism with the dynamism of the forces of production it unleashes (Cohen, 1979). Nevertheless, a danger lies in freezing class formation over time and space. A rigid conception of class structure and class agency ill describes the rapidly evolving contours of capitalist society over the last three or four centuries. The contemporary configuration of classes did not appear at a stroke with the dawn of Mercantilism or the Industrial Revolution, diffusing from Britain along with the expansion of commercial activity and industrialization. Britain is unique in the way its capitalist classes emerged from, and echoed, the previous social order (Anderson, 1964). Nor do international relations between national bourgeoisies (and working classes) suffice to capture the geographical dimension of class, if we are concerned with the simultaneous determination of class structure and class politics within nations and at a global scale (Baran, 1957; Brenner, 1977).

Thus far we have pictured capital and labor fighting over income distribution and political power in a divided and shifting terrain of industrialization. This takes classes as given and asks only about their relative strength and mutual maneuvering.[11] Our vision must be expanded to include the subtle changes in class constitution that take place in and through territorial development. This process of "geographical class formation" has two intertwined dimensions.

First, class coherence may be strengthened or weakened by industrial geography. Consider the force of union tradition in a coal-mining town, or the confusing effects of a plant closure on working class solidarity. The coherence and strength of contending classes is not simply a matter of class politics or class consciousness; it grows out of class practices that are limited and enabled by structural possibilities but not determined by them. Class may therefore increase or decline as a social force relative to other powerful aspects of social life, such as race or nationality. In other words, class struggle is as much over class formation as over positions between pre-existing classes (Przeworski, 1985), and class structure can only define the terrain on which class formation occurs, not determine outcomes (Wright, 1985). In part, this is because most people find themselves in "contradictory class locations," which do not allow a simple sorting of individuals into class positions (Wright, 1976). But, more than this, the formation of classes rests on an ever-shifting basis, which allows for a more creative process of class agency than is captured by the concepts of contradictory class locations or class alliances.[12]

Second, fundamental changes in the character of capitalist production alter the structural foundations of class. While property relations are a necessary condition for class formation (Roemer, 1982; Brenner, 1986), they are not sufficient. Class theorists have wrestled with this quandary at length, and have put forth several candidates for a broader definition

[11] We do not presume, of course, that collective action is always self-conscious or that its effects are necessarily intentional.

[12] On this we differ with Wright in favor of Giddens' structuration approach, in which structure and agency are mutually determining: "Class boundaries [cannot] be settled *in abstracto*; one of the specific aims of class analysis in relation to empirical societies must necessarily be that of determining how strongly, in any given case, the class principle has become established as a mode of structuration" (1980, p. 110).

of class: control of the labor process, control of finance capital, control of mental labor, possession of organizational assets and possession of skills (Braverman, 1974; Carchedi, 1977; Roemer, 1982; Wright, 1985). These can all be reduced to elements in the (increasingly complex) social division of labor and to capacities introduced by the advancing technological and organizational capabilities of capitalist society. They therefore rest, broadly speaking, on the forces of production not the class relations of capitalist society, and cannot be reduced to the latter. Simply expanding the list of class capacities or resources, in this fashion, resolves nothing unless we recognize that class structuring is inextricably bound up with the *development* of capitalism as a system of production, circulation, and exploitation.

The nature of capitalist class relations has thus changed over time. This has led to an unending series of theoretical quarrels: were early merchants true capitalists, bearing with them fully formed capitalism (Wallerstein, 1974; Brenner, 1977)? Did industrial capitalism begin with the putting-out system or the factory (Dobb, 1947; Kriedte et al., 1977)? Did corporate managers force out capitalists as the true leaders of industry (Berle and Means, 1933; Herman, 1981)? Do professionals and technicians constitute a new middle class (Walker, 1979; Przeworski, 1985)? Such quarrels are by no means trivial. They go to the heart of capitalist society and its development.[13] Without presuming to resolve such difficult questions, we can nonetheless say that classes are never finished or fully formed entities. They are forever restructured – though along essentially capitalist lines – in the face of a changing productive base. Moreover, this constitution and reconstitution involves the exercise of power, by which capitalists have been able to maintain their grip on the key domains of economy and civil society in spite of the potential and actual claims of laborers, technocrats, or state executives.

Geographical industrialization figures in class formation, as it has in other processes we have considered. Class formation necessarily takes place on the shifting territory of an unfolding system of production. Class intersects not only the social division of labor, but the spatial division of labor as well.[14] Territorial expansion, differentiation, and upheaval over the course of capitalist development are accompanied by changes in class structuration and class formation that alter the bases of conflicts over wages, work rules, or local government. One must conclude that, on the whole, these patterns of uneven spatial development have had a negative impact on the fortunes of the working class, because they repeatedly

[13] The same problem besets the debate over European feudalism, whose property relations and power bases in land, trade, cities and the state kept evolving, despite certain constant elements (Ashton and Philpin, 1985).

[14] As do gender, race, and other fundamental social divisions and relations of power (cf. Massey, 1984; Montejano, 1987).

undermine place-based experience and solidarity, help bring in new groups of people who lack experience with and commitment to existing class practices, and introduce new industrial divisions of labor that disorient and reconstruct the foundations of class existence. Moreover, this has occurred almost without exception under conditions favorable to the assertion of capitalist power in new places. Geographical industrialization and the inconstant geography of capitalism have therefore been fundamental to the reproduction of capitalist relations of production and the repeated renewal of this most agile of modes of production.

8.3 Conclusion: the geographical constitution of society

The prevailing assumption in the social sciences is that society and economy have geographical outcomes but not geographical foundations. We disagree. In our view the territorial arrangement of activities is central to the broader constitution of any society's economic, social, and political fabric; indeed, societies are shaped only by virtue of their imbrication in territorial formations.

We have argued throughout this book that the specific economic and political forms taken by industries emerge only by means of their geography. Industries do not simply put their imprint on places after having been predetermined in a spaceless environment. In moving from industry, narrowly construed, to political economy and social theory, we have depicted geographical industrialization not only as the motor of economic growth, but as the basis for much of the structuring of class relations. The processes of territorial differentiation, extension and upheaval considered throughout this book are the variegated ground on which class struggle and class formation are played out. The geography of industrialization is therefore fundamental to macroeconomics, macropolitics and to the social life of capitalist societies, and thus to the very shape of capitalism itself. In short, geography is a key to understanding not merely urban or regional outcomes, but processes of economic and social development as a whole. The geographical dimension of capitalist development provides a broad window onto the construction of social history in general. As Soja (1980 p. 211) puts it, "social relations of production are both space-forming and space-contingent." Spatial relations, in other words, are society-forming (Massey, 1984).

In making this ontological claim, we must leave behind the conventional geographer's notion of industrial space in the sense of a geometry of locations, and instead situate industrialization in its territorial milieu. Territories are socially-created congeries of human activities, around which form important geographical boundaries that shape social action and the exercise of power. This extends our claims that industries localize within distinct domains, that technologies and organizational practices can diverge along distinctive national lines, and that local labor markets are bounded by discrete sets of norms, expectations and institutions.

Wherever it exists, capitalism generates powerful impulses to revolutionize the forces and relations of production, the conditions of social life, and the course of human history. These impulses always emerge in concrete forms as a result of real social practices by real people. We have argued in this book that the place-bound mastery of technology, the territorial construction of production systems and politics, and the territorial bounding and differentiation of class relations are key to the dynamics of those processes, and hence to the shape of particular capitalisms. We have also stressed the importance of social-spatial explorations at the social frontiers of growth, opened up by both spatial extension and the persistent uneveness of capitalist territories. But this is not all, for capitalism fundamentally operates via strong competition, and inter-place competition – in a variety of political, economic, and social dimensions – is a vital part of that struggle between new and old in which whole epochs are born, fight for supremacy and pass from the scene.

The particulars of capitalist geography are thus an ontological foundation for capitalism's inconstancy, for its inner dynamics. The inconstant geography of capitalism is the inconstancy of capitalism itself, as a generator of disequilibrium growth and as a mode of production in historical and geographical motion. By grasping the particulars of that inconstant geography, one can get a hold on historical change at a middle level of abstraction and concreteness, between the full sweep of centuries and the endless swirl of everyday events. This implies a more geographical historical materialism. This reintegration of geography into social science at the most basic level is key to understanding how economic growth proceeds, how social inequality is reproduced in new forms, and how people make history under geographical circumstances not of their own choosing.

Bibliography

Abernathy, W., Clark, D. and Kantrow, A. 1983: *Industrial Renaissance: Producing a Competitive Future for America*. New York: Basic Books.

Abramowitz, M. 1956: *Resource and Output Trends in the United States Since 1870*. New York: National Bureau of Economic Research.

Abramowitz, M. 1976: Likenesses and contrasts between the investment boom of the postwar period and earlier periods in relation to long swings in economic growth. In Richards, H. (ed.) *Population, Factory Movements and Economic Development*. Cardiff: University of Wales, 22–49.

Adler, P. 1985: Technology and us. *Socialist Review*, 85:67–98.

Aglietta, M. 1976: *A Theory of Capitalist Regulation*. English edition 1979. London: New Left Books.

Akerlof, G. 1982: Labor contract as partial gift exchange. *Quarterly Journal of Economics*, 97:543–69.

Alchian, A. and Demsetz, H. 1972: Production, information costs, and economic organization. *American Economic Review*, 62:777–95.

Alexander, J. 1954: The basic/non-basic concept of urban economic function. *Economic Geography*, 30:246–61.

Alonso, W. 1980: Five bell shapes in development. *Papers of the Regional Science Association*, 45:5–16.

Alonso, W. and Medrich, E. 1978: Spontaneous growth centers in twentieth century American urbanization. In Bourne, L. and Simmons, J. (eds) *Systems of Cities*. New York: Oxford University Press, 349–61.

Anderson, P. 1964: The origins of the present crisis. *New Left Review*, 23:26–53.

Anderson, P. 1987: The figures of descent. *New Left Review*, 161:20–77.

Armstrong, R. 1972: *The Office Industry: Patterns of Growth and Location*. Cambridge, Mass: MIT Press (for the Regional Plan Association of New York, edited by B. Pushkarev).

Arrow, K. 1962: The economic implications of learning by doing. *Review of Economic Studies*, 29:154–74.

Arrow, K. 1972: Models of job discrimination. In Pascal, A. (ed.) *Racial Discrimination in Economic Life*. Lexington Mass: Lexington Books.

Arthur, W. 1989: Competing technologies, increasing returns and lock-in by historical events. *Economic Journal*, 99 (forthcoming).

Ashton, T. and Philpin, C. (eds) 1985: *The Brenner Debate: Agrarian Class Structure and Economic Development in Pre-Industrial Europe*. New York: Cambridge University Press.

Bagnasco, S. 1977: *Tre Italia: La Problematica dello Svillupo*. Bologna: Il Mulino.

Bailey, E. and Friedlaender, A. 1982: Market structure and multiproduct industries. *Journal of Economic Literature*, 20:1024–48.

Bain, J. 1956: *Barriers to New Competition*. Cambridge Mass: Harvard University Press.

Bairoch, P. 1977: *Taille des Villes, Conditions de Vie, et Développement Économique*. Paris: Editions de l'École des Hautes Études en Sciences Sociales.

Bakis, H. 1977: *IBM: Une Multinationale Régionale*. Grenoble: Presse Universitaire de Grenoble.

Bakis, H. 1980: The communications of larger firms and their implications for the emergence of a new world industrial order: a case study of IBM's global data network. Contributing report, Commission on Industrial Systems, International Geographical Union, 24th Congress.

Baran, B. 1986: The Technological Transformation of White-Collar Work: A Case Study of the Insurance Industry. Doctoral dissertation, Department of City and Regional Planning, University of California, Berkeley.

Baran, P. 1957: *The Political Economy of Growth*. New York: Monthly Review Press.

Baran, P. and Sweezy, P. 1966: *Monopoly Capital*. New York: Monthly Review Press.

Barnes, T. and Sheppard, E. 1985: Technical choice and reswitching in space economies. *Regional Science and Urban Economics*, 14:345–62.

Barnett, H. and Morse, C. 1963: *Scarcity and Growth*. Baltimore: Johns Hopkins Press.

Barsly, C. and Personich, M. 1981: Measuring wage dispersion: pay ranges reflect industry traits. *Monthly Labor Review*, 104(4):35–41.

Beccattini, G. 1987: *Mercato e Forze Locali–Il Distretto Industriale*. Bologna: Il Mulino.

Becker, G. 1964: *Human Capital*. New York: National Bureau of Economic Research and Columbia University Press.

Belil, M. 1985: Subcontracting Networks: An Alternative Form of Capitalist Organization. Master's thesis, Department of Geography, University of California, Berkeley.

Bell, D. 1973: *The Coming of Post-Industrial Society*. New York: Basic Books.

Bell, R. 1972: *Changing Technology and Manpower Requirements in the Engineering Industry*. Brighton: Sussex University Press and Engineering Industry Training Board.

Bellandi, M. 1986: *The Marshallian Industrial District*. Studie Discussioni No. 42. Florence: Dipartimenti di Scienze Economiche, Università degli Studi di Firenze.

Bellofiore, R. 1985: Money and development in Schumpeter. *Review of Radical Political Economy*, 17(1&2):21–40.

Bensel, R. 1983: *Sectionalism in US Politics*. New York: Columbia University Press.

Berger, S. and Piore, M. 1981: *Dualism and Discontinuity in Industrial Societies*. New York: Cambridge University Press.

Berle, A. and Means, G. 1933: *The Modern Corporation and Private Property*. New York: Macmillan.

Berman, M. 1982: *All That Is Solid Melts Into Air*. New York: Simon and Schuster.

Berry, B. 1961: City size distributions and economic development. *Economic Development and Cultural Change*, 9(4):573–87.

Berry, B. 1967: *Geography of Market Centers and Retail Distribution*. Englewood Cliffs, NJ: Prentice-Hall.

Berry, B. 1972: Hierarchical diffusion: the basis of developmental filtering and spread in a system of growth centers. In Hansen, N. (ed.) *Growth Centers in Regional Economic Development*. New York: Free Press, 108–38.

Berry, B. 1976: The counterurbanization process: urban America since 1970. In Berry, B. (ed.) *Urbanization and Counterurbanization*. Beverly Hills: Sage, 17–30.

Beynon, H. 1973: *Working for Ford*. London: Allen Lane.

Blauner, R. 1964: *Alienation and Freedom: The Factory Worker and Industry*. Chicago: University of Chicago Press.

Blaut, J. 1977: Two views of diffusion. *Annals of the Association of American Geographers*, 67(3):343–9.

Bloch, M. 1948: *Feudal Society*. 2 volumes. Chicago: University of Chicago Press.

Block, F. 1977: *The Origins of International Economic Disorder*. Berkeley and Los Angeles: University of California Press.

Bluestone, B. and Harrison, B. 1982: *The Deindustrialization of America*. New York: Basic Books.

Blum, S. 1925: *Labor Economics*. New York: Henry Holt and Co.

Boas, C. 1961: Locational patterns of American automobile assembly plants, 1895–1958. *Economic Geography*, 37:218–30.

Bok, D. 1971: Reflections on the distinctive character of American labor laws. *Harvard Law Review*, 84:1394–1463.

Borts, G. and Stein, J. 1964: *Economic Growth in a Free Market*. New York: Columbia University Press.

Bott, E. 1971: *Family and Social Networks*. New York: Free Press.

Boudeville, J. 1966: *Problems of Regional Economic Planning*. Edinburgh: Edinburgh University Press.

Bourne, L. and Simmons, J. (eds) 1978: *Systems of Cities*. New York: Oxford University Press.

Bowles, S. and Gintis, H. 1976: *Schooling in Capitalist America*. New York: Basic Books.

Bowles, S. and Gintis, H. 1986: *Capitalism and Democracy*. New York: Basic Books.

Bowles, S., Gordon, D. and Weisskopf, T. 1983: *Beyond the Wasteland*. Garden City, NY: Anchor Press/Doubleday.

Braudel, F. 1972: *The Mediterranean and the Mediterranean World in the Age of Philip II*. 2 volumes. New York: Harper and Row.

Braudel, F. 1979: *Civilization and Capitalism, 15th to 18th Century*. 3 volumes. New York: Harper and Row.

Braverman, H. 1974: *Labor and Monopoly Capital*. New York: Monthly Review Press.

Brenner, R. 1977: The origins of capitalist development: a critique of neo-Smithian Marxism. *New Left Review*, 104:25–92.

Brenner, R. 1986: The social basis of economic development. In Roemer, J. (ed.) *Analytical Marxism*. New York: Cambridge University Press, 23–53.

Bright, J. 1958: *Automation and Management*. Boston: Harvard University School of Business Administration.

Brown, L. 1980: *Innovation Diffusion: A New Perspective*. London: Methuen.

Brusco, S. and Sabel, C. 1983: Artisanal production and economic growth. In Wilkinson, F. (ed.) *The Dynamics of Labor Market Segmentation*. London: Academic Press, 99–113.

Buckley, P. and Casson, M. 1976: *The Future of the Multinational Enterprise*. London: Macmillan.

Buder, S. 1967: *Pullman: An Experiment in Industrial Order and Community Planning, 1880–1930*. New York: Oxford University Press.

Burawoy, M. 1976: The functions and reproduction of migrant labor: comparative material from southern Africa and the United States. *American Journal of Sociology*, 81:1050–87.

Burawoy, M. 1979: *Manufacturing Consent*. Chicago: University of Chicago Press.

Burawoy, M. 1981: Terrains of contest: factory and state under capitalism and socialism. *Socialist Review*. 58:83–125.

Burawoy, M. 1985: *The Politics of Production*. London: Verso.

Burns, A. 1934: *Production Trends in the United States Since 1870*. New York: National Bureau of Economic Research.

Burns, A. 1966: *The Pelican History of Greece*. Harmondsworth: Penguin.

Business Week. 1986: Japan, USA. *Special Report*. July 14:45–55.

Business Week. 1987: General Motors: What went wrong? *Special Report*. March 16:102–10.

Calhoun, C. 1982: *The Question of Class Struggle*. Chicago: University of Chicago Press.

Capoglu, G. 1987: Prices, Profits and Financial Structures: a Post-Keynesian Approach. Doctoral dissertation, Department of Economics, University of California, Berkeley.

Carchedi, G. 1977: *On the Economic Identification of Classes*. London: Routledge and Kegan Paul.

Cardellino, J. 1982: Industrial Location: A Case Study of the California Fruit and Vegetable Canning Industry, 1860–1984. Master's thesis, Department of Geography, University of California, Berkeley.

Casetti, E. 1979: The onset of modern economic growth: empirical validation of a catastrophe model. *Papers of the Regional Science Association*, 50:9–20.

Caves, R. 1981: Intra-industry trade and market structure in the industrial countries. *Oxford Economic Papers*, 33:203–33.

Caves, R. 1982: *Multinational Enterprise and Economic Analysis*. Cambridge, UK: Cambridge University Press.

Cebula, R. and Smith, L. 1981: An exploratory empirical note on determinants of interregional cost-of-living differentials in the United States, 1970 and 1975. *Regional Science and Urban Economics*, 11(1):81–5.

Chandler, A. 1962: *Strategy and Structure*. Cambridge, Mass: MIT Press.

Chandler, A. 1977: *The Visible Hand*. Cambridge, Mass: Harvard University Press.

Chenery, H. 1960: Patterns of industrial growth. *American Economic Review*, 50:624–54.

Chinitz, B. 1960: *Freight in the Metropolis*. Cambridge, Mass: Harvard University Press.

Chinitz, B. 1961: Contrasts in agglomeration: New York and Pittsburgh. *American Economic Review, Papers and Proceedings*, 279–89.

Cho, S. 1987: How Cheap is Cheap Labor? The Dilemmas of Export-led Industrialization. Doctoral dissertation, Department of Sociology, University of California, Berkeley.

Christaller, W. 1935: *Central Places in Southern Germany*. English edition 1966. Englewood Cliffs: Prentice-Hall.

Christopherson, S. 1986: *Trends Toward Labor Flexibility in the Reported and Unreported Economy*. Washington, DC: Office of Technology Assessment.

Christopherson, S. and Storper, M. 1988: New forms of labor segmentation and industrial politics in flexible production industries. *Industrial and Labor Relations Review*, 42(3):331–347.

Clark, G. 1981: The employment relation and the spatial division of labor. *Annals of the Association of American Geographers*, 71:412–24.

Clark, G. 1983: Fluctuations and rigidities in local labor markets, part 1: theory and evidence. *Environment and Planning A*, 15:168–85.

Clark, G. 1985: *Judges and the Cities*. Chicago: University of Chicago Press.

Clark, G. 1986: The crisis of the Midwest auto industry. In Scott, A. and Storper, M. (eds) *Production, Work, Territory*. Boston: Allen and Unwin, 127–48.

Clark, G. 1989: *Unions and Communities Under Siege: American Communities and the Crisis of Organized Labor*. Cambridge UK: Cambridge University Press.

Clark, G., Gertler, M. and Whiteman, J. 1986: *Regional Dynamics: Studies in Adjustment Theory*. Boston and London: Allen and Unwin.

Clark, G. and Johnston, K. 1987: The geography of US union elections 1: the crisis of US unions and a critical review of the literature. *Environment and Planning A*, 19:33–57.

Clark, P. 1979: Investment in the 1970s: theory, performance and prediction. *Brookings Papers on Economic Activity*, 1:73–124.

Clarke, I. 1985: *The Spatial Organization of Multinational Corporations*. New York: St Martin's Press.

Coase, R. 1937: The nature of the firm. *Economica*, 4:386–405.

Cohen, G. 1979: *Karl Marx's Theory of History: A Defence*. Princeton, NJ: Princeton University Press.

Cohen, R. 1981: The new international division of labor, multinational corporations and urban hierarchy. In Dear, M. and Scott, A. (eds) *Urbanization and Urban Planning in Capitalist Society*. New York: Methuen, 287–315.

Cohen, S. 1987: A labour process to nowhere? *New Left Review*, 165: 34–51.

Cohen, S. and Zysman, J. 1987: *Manufacturing Matters: The Myth of the Post-Industrial Economy*. New York: Basic Books.

Commoner, B. 1976. *The Poverty of Power: Energy and the Economic Crisis*. New York: Knopf.

Cooke, P. 1986: The changing urban and regional system in the United Kingdom. *Regional Studies*, 20(3):243–51.

Coriat, B. 1979: *L'Atelier et le Chronomètre*. Paris: Christian Bourgois.

Cripps, T. and Tarling, R. 1973: *Growth in Advanced Capitalist Countries, 1950–1970*. Cambridge, UK: Cambridge University Press.

Crompton, R. and Jones, G. 1984: *White Collar Proletariat: Deskilling and Gender in Clerical Work*. Philadelphia: Temple University Press.

Cumings, B. 1984: The origins and development of the Northeast Asian political economy: industrial sectors, product cycles, and political consequences. *International Organization*, 38(1):1–40.

Cunningham, 1951: *The Aircraft Industry: A Study in Industrial Location*. Los Angeles: Morrison.

Cusumano, M. 1985: *The Japanese Automobile Industry*. Cambridge, Mass: Harvard University Press.

Czamanski, S. and Ablas, L. 1979: Identification of industrial clusters and complexes: a comparison of methods and findings. *Urban Studies*, 16(1):61–80.

Danhof, C. 1969: *Change in Agriculture: The Northern US, 1820–1870*. Cambridge, Mass: Harvard University Press.

Daniels, P. (ed.) 1979: *Spatial Patterns of Office Growth and Location*. New York: John Wiley.

Daniels, P. 1982: *Service Industries: Growth and Location*. Cambridge, UK: Cambridge University Press.

David, P. 1975: *Technical Choice, Innovation and Economic Growth*. New York: Cambridge University Press.

Davis, M. 1986: *Prisoners of the American Dream*. London: Verso.

Davis, M. 1987: *Chinatown*, part two? The "internationalization" of downtown Los Angeles. *New Left Review*, 164:65–86.

Dawley, A. 1976: *Class and Community: The Industrial Revolution in Lynn*. Cambridge, Mass: Harvard University Press.

DeBresson, C. 1989: Technological clusters: poles of development. *World Development* (forthcoming).

DeBresson, C. and Townsend, J. 1978: Notes on the inter-industrial flow of technology in post-war Britain. *Research Policy*, 7:48–60.

DeGeer, S. 1927: The American manufacturing belt. *Geografisker Annaler*, 9(4):233–59.

DeJanvry, A. 1981: *The Agrarian Question and Reformism in Latin America*. Baltimore: Johns Hopkins Press.

DeVries, J. 1984: *European Urbanization, 1500–1800*. Cambridge, Mass: Harvard University Press.

DeVroey, M. 1984: A regulation approach interpretation of the contemporary crisis. *Capital and Class*, 23:45–66.

DeVyver, F. 1951: The labor factor in the industrial development of the South. *Southern Economic Journal*, 18:189–205.

Dean, J. 1950: Pricing policies for new products. *Harvard Business Review*, 28:45–53.

Deane, P. 1965: *The First Industrial Revolution*. Cambridge, UK: Cambridge University Press.

Devine, J. 1980: Overinvestment and Cyclic Economic Crises. Doctoral dissertation, Department of Economics, University of California, Berkeley.

Dicken, P. 1986: *Global Shift*. London: Harper and Row.

Dicken, P. and Lloyd, P. 1977: *Location in Space*. 2nd ed. New York: Harper and Row.

Dickens, W. and Lang, K. 1985: A test of dual labor market theory. *American Economic Review*, 75:792–805.

Di Tella, G. 1982: The economics of the frontier. In Kindleberger, C. and Di Tella, G. (eds.) *Economics in the Long View*. New York: New York University Press, 210–27.

Dobb, M. 1947: *Studies in the Development of Capitalism*. New York: International Publishers.

Dollar, D. 1986: Technological innovation, capital mobility and the product cycle in north–south trade. *American Economic Review*, 76:177–90.

Domhoff, W. 1979: *The Powers That Be*. New York: Vintage.

Dore, R. 1987: *Taking Japan Seriously*. Stanford: Stanford University Press.

Dosi, G. 1984: *Technical Change and Industrial Transformation*. New York: St Martin's Press.

Dumenil, G., Glick, M. and Rangel, J. 1987: The rate of profit in the United States. *Cambridge Journal of Economics*, 11(4):331–59.

Dunlop, J. 1944: *Wage Determination Under Trade Unions*. New York: Kelley.

Dunlop, J. 1948: *Collective Bargaining*. Chicago: R. D. Irwin.

Dunning, J. 1979: Explaining changing patterns of international production: in defence of the eclectic theory. *Oxford Bulletin of Economics and Statistics*, 41:269–95.

Dunning, J. 1981: *International Production and the Multinational Enterprise*. London: Allen and Unwin.

Durand, C. and Durand, M. 1971: *De l'O.S. à l'Ingénieur: Carrière ou Classe Sociale*. Paris: Les Éditions Ouvrières.

Dworkin, R. 1985: *A Matter of Principle*. Cambridge, Mass: Harvard University Press.

Ebel, K. 1985: Social and labor implications of flexible manufacturing systems. *International Labour Review*, 124(2):133–45.

Edwards, R. 1979: *Contested Terrain*. New York: Basic Books.

Edwards, R., Reich, M. and Gordon, D. (eds) 1975: *Labor Market Segmentation*. Lexington, Mass: D.C. Heath.

El Shaks, S. 1972: Development, primacy, and systems of cities. *Journal of Developing Areas*, 7:11–36.

Erickson, R. 1974: The regional impact of growth firms: the case of Boeing. *Land Economics*, 50(2):127–36.

Erickson, R. 1975: The spatial pattern of income generation in lead firm, growth area, and linkage systems. *Economic Geography*, 51(1):17–26.

Erickson, R. and Leinbach, T. 1979: Characteristics of branch plants attracted to nonmetropolitan areas. In Lonsdale, R. and Seyler, H. (eds) *Nonmetropolitan Industrialization*. Washington, DC: V. H. Winston, 57–78.

Estall, R. 1985: Stock control in manufacturing: the just-in-time system and its locational implications. *Area*, 17(2):129–33.

Ethier, W. 1982: National and international returns to scale in the modern theory of international trade. *American Economic Review*, 72(3):389–404.

Farjoun, E. and Machover, M. 1983: *Laws of Chaos*. London: Verso.

Feller, I. 1975: Invention, diffusion and industrial location. In Collins, L. and Walker, D. (eds) *Locational Dynamics of Manufacturing Activity*. London: Wiley, 83–108.

Fine, B. and Harris, L. 1979: *Re-reading Capital*. London: Macmillan.

Fischer, C. 1976: *The Urban Experience*. New York: Harcourt Brace Jovanovich.

Fishlow, A. 1965: *American Railroads and the Transformation of the Ante-Bellum Economy*. Cambridge, Mass: Harvard University Press.

Fitzsimmons, M. 1986: The new industrial agriculture: the regional integration of specialty crop production. *Economic Geography*, 62(4):334–52.

Florida, R. and Kenney, M. 1988: Venture capital, high technology and regional development. *Regional Studies*, 22(1):33–48.

Florida, R. and Kenney, M. 1989: *The Breakthrough Economy*. New York: Basic Books.

Foster, J. 1974: *Class Struggle and the Industrial Revolution*. London: Weidenfeld and Nicholson.

Foster-Carter, A. 1985: Korea and dependency theory. *Monthly Review*, 37(5):27–34.

Frank, A. 1969: *Capitalism and Underdevelopment in Latin America*. New York: Monthly Review Press.

Freedman, M. 1976: *Labor Markets: Segments and Shelters*. Montclair, NJ: Allenheld, Osmun.

Freeman, C. 1982: *The Economics of Industrial Innovation*. 2nd ed. London: Frances Pinter.

Freeman, C., Clark, J. and Soete, L. 1982: *Unemployment and Technical Innovation*. Westport, Conn: Greenwood Press.

Freeman, R. 1978: Job satisfaction as an economic variable. *American Economic Review*, 68(2):135–41.

Freeman, R. and Medoff, J. 1984: *What Do Unions Do?* New York: Basic Books.

Freiberger, P. and Swaine, M. 1984: *Fire in the Valley: The Making of the Personal Computer*. Berkeley: Osborne/McGraw Hill.

Friedland, R. and Bielby, W. 1981: The power of business in the city. In Clark, T. (ed.) *Urban Policy Analysis*. Beverly Hills: Sage, 133–51.

Friedman, A. 1977: *Industry and Labour*. London: Macmillan.

Froebel, F., Heinrichs, J. and Kreye, O. 1977: *The New International Division of Labor*. English edition 1980. Cambridge UK: Cambridge University Press.

Fuchs, V. 1962: *The Changing Location of Manufacturing in the U.S. Since 1929*. New Haven: Yale University Press.

Gaffney, M. 1962: Land and rent in welfare economics. In Clawson, M., Harris, C. and Ackerman, W. (eds) *Land Economics Research*. Baltimore: Johns Hopkins Press, 41–67.

Galambos, L. 1966: *Competition and Cooperation: The Emergence of a National Trade Association*. Baltimore: Johns Hopkins Press.

Galbraith, J. 1967: *The New Industrial State*. Boston: Houghton Mifflin.

Gallie, D. 1979: *In Search of the New Working Class*. New York: Cambridge University Press.

Gerschenkron, A. 1962: *Economic Backwardness in Historical Perspective*. Cambridge, Mass: Harvard Belknap Press.

Giddens, A. 1980: *The Class Structure of the Advanced Societies*. 2nd ed. London: Hutchinson.

Giddens, A. 1984: *The Constitution of Society*. Oxford: Polity Press and Berkeley: University of California Press.

Gintis, H. and Bowles, S. 1981: Structure and practice in the labor theory of value. *Review of Radical Political Economics*, 12(4):1–25.

Glasmeier, A. 1985: The Structure, Location and Role of High Technology Industries in US Regional Development. Doctoral dissertation, Department of City and Regional Planning, University of California, Berkeley.

Gold, B. 1964: Industry growth patterns: theory and empirical results. *Journal of Industrial Economics*, 13:53–73.

Gold, B. (ed.) 1977: *Research, Technological Change and Economic Analysis*. Lexington: Lexington Books.

Gold, D. 1987. Explaining New York's metropolitan dominance. Paper delivered at the Social Science Research Council workshop on the Dual City, New York.

Gordon, D. 1977: Class struggle and the stages of American urban development. In Perry, D. and Watkins, A. (eds) *The Rise of the Sunbelt Cities*. Beverly Hills: Sage, 55–82.

Gould, S. 1977: *Ever Since Darwin*. New York: Norton.

Gramsci, A. 1934: *Prison Notebooks*. English edition 1971. New York: International Publishers.

Gregory, D. 1978: *Science, Ideology and Human Geography*. London: Hutchinson.

Gregory, D. and Urry, J. 1985: *Social Structure and Spatial Relations*. London: Macmillan.

Griliches, Z. 1957: Hybrid corn: an exploration in the economics of technological change. *Econometrica*, 25:501–22.

Griliches, Z. 1969: Capital–skill complementarity. *Review of Economics and Statistics*, 51:465–68.

Habakkuk, H. 1962: *American and British Technology in the 19th Century*. Cambridge, UK: Cambridge University Press.

Hägerstrand, T. 1953: *Innovation Diffusion as a Spatial Process*. English edition 1967. Chicago: University of Chicago Press.

Hall, P. 1962: *The Industries of London Since 1861*. London: Hutchinson.

Hall, P. 1964: Industrial London: a general view. In Coppock, J. and Prince, H. (eds) *Greater London*. London: Faber and Faber, 225–46.

Hall, P., Breheny, M., McQuaid, R. and Hart, D. 1987: *Western Sunrise: The Genesis and Growth of Britain's High Tech Corridor*. London: Allen and Unwin.

Hall, P. and Markusen, A. (eds) 1985: *Silicon Landscapes*. Boston and London: Allen and Unwin.

Hamer, A. 1973: *Industrial Exodus from the Central City*. Lexington, Mass: Lexington Books.

Hamermesh, D. 1975: Interdependence in the labor market. *Economica*, 42:420–9.

Hamilton, F. (ed.) 1974: *Spatial Perspectives on Industrial Organization and Decisionmaking*. London: Wiley.

Hansen, N. 1979: The new international division of labor and manufacturing decentralization in the United States. *Review of Regional Studies*, 9:1–11.

Hansen, N. 1981: *The Border Economy: Regional Development in the Southwest*. Austin: University of Texas Press.

Hansen, N. 1982: The South and export-oriented industrialization. Paper presented to Regional Science Association, North American Meetings, Denver, Colo.

Harcourt, G. 1972: *Some Cambridge Controversies in the Theory of Capital*. Cambridge, UK: Cambridge University Press.

Hareven, T. 1982: *Family Time and Industrial Time*. New York: Cambridge University Press.

Harrigan, K. 1985: *Strategies for Joint Ventures*. Lexington, Mass: Lexington Books.

Harris, D. 1978: *Capital Accumulation and Income Distribution*. Stanford: Stanford University Press.

Harris, D. 1982: Structural change and economic growth: a review article. *Contributions to Political Economy*, 1:25–45.

Harris, D. 1983: Accumulation of capital and the rate of profit in Marxian theory. *Cambridge Journal of Economics*, 7:311–30.

Harris, D. 1986: Are there macroeconomic laws? The law of the falling rate of profit reconsidered. In Wegener, H. and Drukker, J. (eds) *The Economic Law of Motion of Modern Society*. New York: Cambridge University Press, 49–63.

Harrison, B. 1984: Regional restructuring and 'good business climates.' In Sawers, L. and Tabb, W. (eds) *Sunbelt/Snowbelt: Urban Development and Regional Restructuring*. New York: Oxford University Press, 48–96.

Harrison, B. and Bluestone, B. 1988: *The Great U-Turn : Corporate Restructuring, Laissez Faire and the Rise of Inequality in America*. New York: Basic Books.

Harrod, R. 1973: *Economic Dynamics*. New York: St Martin's Press.

Hartman, C. 1984: *The Transformation of San Francisco*. Totowa, NJ: Rowman and Allenheld.

Hartman, H. 1979: Capitalism, patriarchy and job segregation by sex. In Eisenstein, Z. (ed.) *Capitalist Patriarchy and the Case for Socialist Feminism*. New York: Monthly Review Press, 206–47.

Harvey, D. 1973: *Social Justice and the City*. Baltimore: Johns Hopkins Press; Oxford: Basil Blackwell (1988).

Harvey, D. 1975: The geography of capital accumulation. *Antipode*, 7(2):9–21.

Harvey, D. 1982: *The Limits to Capital*. Oxford: Basil Blackwell; Chicago: University of Chicago Press.

Harvey, D. 1984: On the history and present condition of geography: an historical materialist manifesto. *Professional Geographer*, 36(1):1–10.

Harvey, D. 1985: *The Urbanization of Capital*. Oxford: Basil Blackwell; Baltimore: Johns Hopkins Press.

Hayami, Y. and Ruttan, V. 1971: *Agricultural Development*. Baltimore: Johns Hopkins Press.

Hayter, R. and Watts, H. 1983: The geography of enterprise: a reappraisal. *Progress in Human Geography*, 7:157–81.

Heiman, M. 1988: *The Quiet Evolution: Power, Planning and Profits in New York*. New York: Praeger.

Hekman, J. 1980a: The product cycle and New England textiles. *Quarterly Journal of Economics*, 94(4):697–717.

Hekman, J. 1980b: Can New England hold onto its high-technology industry? *New England Economic Review*, March–April:35–44.

Hekman, J. and Strong, J. 1981: The evolution of New England industry. *New England Economic Review*, March–April:35–46.

Hennart, J-F. 1982: *A Theory of Multinational Enterprise*. Ann Arbor: University of Michigan.

Herman, E. 1981: *Corporate Control, Corporate Power*. New York: Cambridge University Press.

Hill, R. and Feagin, J. 1987: Detroit and Houston: two cities in global perspective. In Smith, M. and Feagin, J. (eds) *The Capitalist City*. Oxford: Basil Blackwell, 155–77.

Hirsch, S. 1967: *Location of Industry and International Competitiveness*. Oxford: Clarendon Press.

Hirsch, S. E. 1978: *Roots of the American Working Class: A Study of 19th Century Newark*. Philadelphia: University of Pennsylvania Press.

Hirschhorn, L. 1984: *Beyond Mechanization*. Cambridge, Mass: MIT Press.

Hirschman, A. 1958: *The Strategy of Economic Development*. New Haven: Yale University Press.

Hodgson, G. 1988: *Economics and Institutions*. Philadelphia: University of Pennsylvania Press.

Holland, S. 1976: *Capital Versus the Regions*. London: Macmillan.

Holmes, J. 1986: The organizational and locational structure of production subcontracting. In Scott, A. and Storper, M. (eds) *Production, Work, Territory*. Boston and London: Allen and Unwin, 80–106.

Hoover, E. 1948: *The Location of Economic Activity*. New York: McGraw-Hill.

Hoover, E. and Vernon, R. 1959: *Anatomy of a Metropolis*. Cambridge, Mass: Harvard University Press.

Hossfeld, K. 1988: Divisions of Labor, Divisions of Lives: Immigrant Women Workers in Silicon Valley. Doctoral dissertation, Department of Sociology, University of California, Santa Cruz.

Hounshell, D. 1984: *From the American System to Mass Production, 1800–1932*. Baltimore: Johns Hopkins Press.

Hudson, J. 1969: Diffusion in a central place system. *Geographical Analysis*, 1:45–58.

Hudson, R., Anderson, J., Duncan, S. and London, R. 1983: *Redundant Spaces in Cities and Regions: Studies in Industrial Decline and Social Change*. New York: Academic Press.

Hunt, E. and Schwartz, J. (eds) 1972: *A Critique of Economic Theory*. Baltimore: Penguin.

Hymer, S. 1972: The multinational corporation and the law of uneven development. In Bhagwai, J. (ed.) *Economics and World Order*. New York: Free Press, 113–40.

Ikeda, M. 1979: The subcontracting system in the Japanese electronic industry. *Engineering Industries of Japan*, 19:43–71.

Isard, W. 1956: *Location and Space Economy*. New York: Wiley.

Isard, W. 1969: *General Theory: Social, Political, Economic, and Regional*. Cambridge, Mass: MIT Press.

Jacobsson, S. 1985: Technical change and industrial policy: the case of Korea and Taiwan. *World Development*, 13(3):353–70.

Jayet, H. 1983: Chômer plus souvent en région urbaine, plus longtemps en région rurale. *Économie et Statistique*, 153:47–57.

Jenkins, R. 1984: Divisions over the international division of labor. *Capital and Class*, 22:28–58.

Jewkes, J., Sawers, D. and Stillerman, R. 1959: *The Sources of Invention*. New York: St Martin's Press.

Johnson, C. 1982: *MITI and the Japanese Miracle*. Stanford: Stanford University Press.

Johnston, K. 1986: Judicial adjudication and the spatial structure of production: two decisions by the National Labor Relations Board. *Environment and Planning A*, 18:27–39.

Jorgenson, D. 1971: Econometric studies of investment behavior: a survey. *Journal of Economic Literature*, 9:1111–47.

Jusenius, L. and Ledebur, L. 1978: *Documenting the Decline of the North*. Washington, DC: Economic Development Administration, US Department of Commerce.

Kafkalas, G. 1985: Location of production and forms of spatial integration. *International Journal of Urban and Regional Research*, 9(2):233–53.

Kain, J. 1968: The distribution and movement of jobs and industry. In Wilson, J. (ed.) *The Metropolitan Enigma*. Cambridge, Mass: Harvard University Press, 1–43.

Kaldor, N. 1956: Alternative theories of distribution. *Review of Economic Studies*, 23(2):143–56.

Kaldor, N. 1970: The case for regional policies. *Scottish Journal of Political Economy*, 17:337–47.

Kaldor, N. 1972: The irrelevance of equilibrium economics. *Economic Journal*, 82:1237–55.

Kalecki, M. 1954: *Theory of Economic Dynamics*. London: Allen and Unwin.

Kalecki, M. 1971: *Selected Essays on the Dynamics of the Capitalist Economy, 1922–1970*. Cambridge, UK: Cambridge University Press.

Kalleberg, A. and Griffin, L. 1978: Positional sources of inequality in job satisfaction. *Sociology of Work and Occupations*, 5:371–401.

Kaplinsky, R. 1984: *Automation*. Harlow, UK: Longman.

Katz, J. 1983: Domestic technological innovations and dynamic comparative advantages. Unpublished paper, Economic Commission for Latin America, Buenos Aires.

Kazin, M. 1987: *Barons of Labor: The San Francisco Building Trades and Union Power in the Progressive Era*. Urbana and Chicago: University of Illinois Press.

Kelley, A. and Williamson, J. 1984: *What Drives Third World City Growth? A Dynamic General Equilibrium Model*. Princeton: Princeton University Press.

Kelly, J. 1985: Management's redesign of work: labour process, labour markets, and product markets. In Knights, D., Wilmott, H. and Colinson, D. (eds) *Job Redesign: Critical Perspectives on the Labour Process*. Aldershot: Gower, 30–51.

Kendrick, J. 1961: *Productivity Trends in the United States*. Princeton: Princeton University Press for National Bureau of Economic Research.

Kendrick, J. 1973: *Post-War Productivity Trends in the United States, 1948–69*. New York: National Bureau of Economic Research.

Kennedy, C. and Thirlwall, A. 1972: Surveys in applied economics: technical progress. *Economic Journal*, 82:11–72.

Keynes, J. 1936: *The General Theory of Employment, Interest and Money*. London: Macmillan.

Klein, B. 1984: Contract costs and administered wages. *American Economic Review*, 74(2):332–44.

Korpi, W. and Shalev, M. 1979: Strikes, industrial relations and class conflict in capitalist societies. *British Journal of Sociology*, 30:164–87.

Korpi, W. and Shalev, M. 1980: Strikes, power and politics in the western nations, 1900–1976. *Political Power and Social Theory*, 1:301–34.

Koutsoyiannis, A. 1984: Goals of oligopolistic firms: an empirical test of competing theories. *Southern Economic Journal*, 51:540–67.

Kraushaar, R. and Feldman, M. 1988: Industrial restructuring and the limits of industry data: examples from western New York. *Regional Studies*, 23.

Kriedte, P., Medich, H. and Schlumbohm, J. 1977: *Industrialization Before Industrialization*. English edition 1981. New York: Cambridge University Press.

Krumme, G. 1969: Toward a geography of enterprise. *Economic Geography*, 45(1):30–40.

Krumme, G. and Hayter, R. 1975: Implications of corporate strategies and product cycle adjustments for regional employment changes. In Collins, L. and Walker, D. (eds) *Locational Dynamics of Manufacturing Activity*. London: Wiley, 325–56.

Kuznets, S. 1930: *Secular Movements in Production and Prices*. Boston: Houghton Mifflin.

Kuznets, S. 1957: Quantitative aspects of the economic growth of nations: industrial distribution of the national product and labor force. *Economic Development and Cultural Change*, 5(4)(suppl):3–111.

Kuznets, S. 1966: *Modern Economic Growth: Rate, Structure, and Spread*. New Haven: Yale University Press.

Lampard, E. 1955: The history of cities in the economically advanced areas. *Economic Development and Cultural Change*, 3(2):81–102.

Landes, D. 1970: *The Unbound Prometheus*. London: Cambridge University Press.

Langenbruch, J. 1981: Os encabeçamentos das armaduras nacionais: uma revisao. *Geografia*, 11(2):1–104.

Lefebvre, H. 1974: *The Production of Space*. English edition 1989. Oxford: Basil Blackwell.

Lenin, V. 1917: Imperialism: the highest stage of capitalism. In *Selected Works*. 3 vols. 1970. Moscow: Progress Publishers, I:667–768.

Leontief, W. 1956: Factor proportions and the structure of American trade: further theoretical and empirical analysis. *Review of Economics and Statistics*, 38:386–407.

Lever, W. 1975: Manufacturing decentralization and shifts in factor costs and external economies. In Collins, L. and Walker, D. (eds) *Locational Dynamics of Manufacturing Activity*. London: Wiley, 295–324.

Levitt, T. 1965: Exploit the product life cycle. *Harvard Business Review*, 43:81–94.

Lindstrom, D. 1978: *Economic Development of the Philadelphia Region, 1810–50*. New York: Columbia University Press.

Lipietz, A. 1980: Interregional polarization and the tertiarisation of society. *Papers of the Regional Science Association*, 44:3–17.

Lipietz, A. 1986: New tendencies in the international division of labor: regimes of accumulation and modes of regulation. In Scott, A. and Storper, M. (eds) *Production, Work, Territory*. Boston: Allen and Unwin, 16–39.

Lipietz, A. 1987: *Mirages and Miracles: The Crises of Global Fordism*. London: Verso.

Littler, C. 1982: *The Development of the Labour Process in Capitalist Societies*. London: Heinemann.

Littler, C. 1985: Taylorism, Fordism and job design. In Knights, D., Willmott, H. and Collinson, D. (eds) *Job Redesign*. Aldershot, UK: Gower, 1–9.

Lloyd, P. and Dicken, P. 1977: *Location in Space*. 2nd ed. New York: Harper and Row.

Locke, R. 1987: Flexible integration in the industrial restructuring of Fiat. Paper presented at the Conference on Industrial Relations in the Era of Flexibility: Adjustments to a Changing Competitive Environment, Cambridge, Mass.

Loertscher, R. and Wolter, F. 1980: Determinants of intra-industry trade. *Welwirtschafliches Archiv*, 116:280–93.

Logan, J. and Molotch, H. 1986: *Urban Fortunes: The Political Economy of Place*. Los Angeles: University of California Press.

Losch, A. 1944: *The Economics of Location*. English edition 1967. New Haven: Yale University Press.

Low-Beer, J. 1978: *Protest and Participation*. Cambridge, UK: Cambridge University Press.

Luxemburg, R. 1913: *The Accumulation of Capital*. English edition 1968. New York: Monthly Review Press.

Lynch, A. 1973: Environment and labor quality take top priority in site selection. *Industrial Development*, 142:13–15.

Macdonald, S. 1983: Technology beyond machines. In Macdonald, S., Lamberton, D. and Mandeville, T. (eds) *The Trouble with Technology*. London: Frances Pinter, 26–36.

Mair, A., Florida R. and Kenney, M. 1988: The new geography of automobile production: Japanese transplants in North America. Working paper 88–45, School of Urban and Public Affairs, Carnegie–Mellon University, Pittsburgh.

Malecki, E. 1981: Science, technology and regional economic development: review and prospects. *Research Policy*, 10(1):312–34.

Malecki, E. 1983: Technology and regional development: a survey. *International Regional Science Review*, 8(2):89–126.

Malecki, E. 1985: Industrial location and corporate organization in high technology industries. *Economic Geography*, 61(4):345–69.

Malecki, E. 1986: Research and development and the geography of high-technology complexes. In Rees, J. (ed.) *Technology, Regions and Policy*. Totowa, NJ: Rowman and Littlefield, 51–73.

Mandel, E. 1975: *Late Capitalism*. London: New Left Books.

Mandel, E. and Freeman, A. 1984: *Ricardo, Marx, Sraffa*. London: Verso.

Mann, E. 1988: *Taking on General Motors*. Los Angeles: Institute of Industrial Relations, University of California, Los Angeles.

Mansfield, E. 1968: *The Economics of Technological Change*. New York: Norton.

Mansfield, E. 1972: Contribution of R and D to economic growth in the United States. *Science*, 175:477–86.

Mansfield, E. 1981: Composition of R and D expenditures: relationship to size of firm, concentration and innovative output. *Review of Economics and Statistics*, 63:610–15.

Mantoux, P. 1961: *The Industrial Revolution in the Eighteenth Century: An Outline of the Beginnings of the Modern Factory System*. London: Cape.

Marglin, S. 1974: What do bosses do? *Review of Radical Political Economy*, 6(2):60–92.

Markusen, A. 1985: *Profit Cycles, Oligopoly, and Regional Development*. Cambridge, Mass: MIT Press.

Markusen, A., Hall, P. and Glasmeier, A. 1986: *High-Tech America: The What, How, Where and Why of the Sunrise Industries*. Boston: Allen and Unwin.

Marris, S. and Wood, A. (eds) 1971. *The Corporate Economy*. Cambridge, Mass: Harvard University Press.

Marshall, A. 1900: *Elements of Economics of Industry*. New York: Macmillan.

Martin, R. and Rowthorn, B. (eds) 1986: *The Geography of Deindustrialization*. London: Macmillan.

Martinelli, F. 1986: Producer Services in a Dependent Economy. Doctoral dissertation, Department of City and Regional Planning, University of California, Berkeley.

Marx, K. 1867: *Capital, Volume I*. English edition 1967. New York: International Publishers.

Marx, K. 1893: *Capital, Volume II*. English edition 1967. New York: International Publishers.

Marx, K. 1894: *Capital, Volume III*. English edition 1967. New York: International Publishers.

Marx, K. and Engels, F. 1848: *Manifesto of the Communist Party*. English edition 1952. Moscow: Progress Publishers.

Massey, D. 1979: In what sense a regional problem? *Regional Studies*, 13(2): 233–43.

Massey, D. 1984: *Spatial Divisions of Labor: Social Structures and the Geography of Production*. London: Macmillan.

Massey, D. and Meegan, R. 1978: Industrial restructuring versus the cities. *Urban Studies*, 15:273–88.

Massey, D. and Meegan, R. 1982: *The Anatomy of Job Loss*. London: Methuen.

McCombie, J. 1981: Kaldor's laws in retrospect. *Journal of Post-Keynesian Economics*, 5:414–29.

McDermott, P. and Taylor, M. 1982: *Industrial Organization and Location*. Cambridge, UK: Cambridge University Press.

McDonald, I. and Solow, R. 1985: Wages and employment in a segmented labor market. *Quarterly Journal of Economics*, 100(4):1115–42.

McLaughlin, G. and Robock, S. 1949: *Why Industry Moves South*. Washington, DC: Committee of the South, National Planning Association.

McMahon, P. and Tschetter, J. 1986: The declining middle class: a further analysis. *Monthly Labor Review*, 109(9):22–7.

McWilliams, C. 1949: *California: the Great Exception*. Reprint 1976. Santa Barbara: Peregrine Smith.

Mensch, G. 1979: *Stalemate in Technology*. Cambridge, Mass: Ballinger.

Meyer, D. 1983: Emergence of the American manufacturing belt: an interpretation. *Journal of Historical Geography*, 9(2):145–74.

Mills, E. 1972: *Studies in the Structure of the Urban Economy*. Baltimore: Johns Hopkins Press.

Mollenkopf, J. 1983: *The Contested City*. Princeton, NJ: Princeton University Press.

Montejano, D. 1987: *Anglos and Mexicans in the Making of Texas, 1836–1986*. Austin: University of Texas Press.

Montgomery, D. 1967: *Beyond Equality*. New York: Alfred Knopf.

Montgomery, D. 1979: *Workers' Control in America*. New York: Cambridge University Press.

Morales, R. 1984: Transitional labor: undocumented workers in the Los Angeles automobile industry. *International Migration Review*, 17(4):570–96.

Morgan, K. 1985: Regional regeneration in Britain: the territorial imperative and the conservative state. *Political Studies*, 33:560–77.

Morgan, K. and Sayer, A. 1988: *Microcircuits of Capital*. Oxford: Polity Press.

Moriarty, B. 1977: Manufacturing wage rates, plant location, and plant location policies. *Popular Government*, 42:48–53.

Morishima, M. 1982: *Why has Japan Succeeded?* Cambridge, UK: Cambridge University Press.

Moses, L. and Williamson, H. 1967: The location of economic activity in cities. *American Economic Review, Papers and Proceedings*, 57:211–22.

Moulaert, F. 1987: An institutional revisit to the Storper–Walker theory of labour. *International Journal of Urban and Regional Research*, 11(3):309–30.

Muller, E. 1977: Regional urbanization and the selective growth of towns in North American regions. *Journal of Historical Geography*, 3(1):21–39.

Muller, E. and Groves, P. 1979: The emergence of industrial districts in mid-nineteenth century Baltimore. *Geographical Review*, 69:157–78.

Mumford, L. 1961: *The City in History*. New York: Harcourt.

Murray, F. 1983: The decentralization of production: the decline of the mass collective worker. *Capital and Class*, 19:74–99.

Musson, A. and Robinson, I. 1969: *Science and Technology in the Industrial Revolution*. Manchester: Manchester University Press.

Myrdal, G. 1957: *Economic Theory and the Underdeveloped Regions*. London: Duckworth.

National Resources Committee. 1937: *Our Cities: Their Role in the National Economy*. Report of the Urbanism Committee. Washington, DC: US Government Printing Office.

National Resources Planning Board. 1943: *Industrial Location and National Resources*. Washington, DC: US Government Printing Office.

Nell, E. 1972: Economics: the revival of political economy. In Blackburn, R. (ed.) *Ideology in Social Science*. London:Fontana/Collins, 76–95.

Nell, E. 1983: Review of Murray Milgate's *Capital and Employment*. *Contributions to Political Economy*, 2:109–15.

Nelson, D. 1975: *Workers and Managers: Origins of the New Factory System in the United States, 1880–1920*. Madison: University of Wisconsin Press.

Nelson, K. 1986: Labor demand, labor supply, and the suburbanization of low-wage office work. In Scott, A. and Storper, M. (eds) *Production, Work, Territory*. Boston: Allen and Unwin, 149–71.

Nelson, R. and Winter, S. 1982: *An Evolutionary Theory of Economic Change*. Cambridge, Mass: Harvard University Press.

Nichols, T. and Beynon, H. 1977: *Living with Capitalism*. London: Routledge and Kegan Paul.

Noble, D. 1986: *Forces of Production: A Social History of Industrial Automation*. New York: Oxford University Press.

North, D. 1955: Location theory and regional economic growth. *Journal of Political Economy*, 63:243–58.

North, D. 1961: *The Economic Growth of the United States, 1790–1860*. Englewood Cliffs: Prentice-Hall.

Norton, R. 1986: Industrial policy and American renewal. *Journal of Economic Literature*, 24:1–40.

Norton, R. and Rees, J. 1979: The product cycle and the spatial decentralization of American manufacturing. *Regional Studies*, 13:141–51.

Noyelle, T. and Stanback, T. 1984: *The Economic Transformation of American Cities*. Totowa, NJ: Rowman and Allenheld.

Ohlin, B. 1939: *Interregional and International Trade*. Cambridge, Mass: Harvard University Press.

Okimoto, D., Sugano, T. and Franklin, B. 1984: *Competitive Edge: The Semiconductor Industry in the U.S. and Japan*. Stanford: Stanford University Press.

Okishio, N. 1961: Technical change and the rate of profit. *Kobe University Economic Review*, 7:85–99.

Ong, A. 1987: *Spirits of Resistance and Capitalist Discipline: Factory Women in Malaysia*. Albany: State University of New York Press.

Organization for Economic Cooperation and Development. 1986: *Flexibility in the Labour Market: The Current Debate*. Paris: OECD.

Panzar, J. and Willig, R. 1977: Economies of scale in multioutput production. *Quarterly Journal of Economics*, 91(3):431–93.

Panzar, J. and Willig, R. 1981: Economies of scope. *American Economic Review*, 71(2):268–72.

Pasinetti, L. 1977: *Lectures on the Theory of Production*. Cambridge, UK: Cambridge University Press.

Pasinetti, L. 1981: *Structural Change and Economic Growth*. New York: Cambridge University Press.

Peet, R. 1984: Class struggle, the relocation of employment, and economic crisis. *Science and Society*, 48(1):38–51.

Peet, R. (ed.) 1987: *International Capitalism and Industrial Restructuring*. Boston: Allen and Unwin.

Perloff, H. and Dodds, V. 1963: *How a Region Grows*. New York: Committee for Economic Development.

Perloff, H., Lampard, E. and Muth, R. 1960: *Regions, Resources, and Economic Growth*. Baltimore: Johns Hopkins Press.

Perroux, F. 1950: Economic space: theory and applications. *Quarterly Journal of Economics*, 64(1):89–104.

Perry, D. 1987: The politics of dependency in deindustrializing America: the case of Buffalo. In Smith, M. and Feagin, J. (eds) *The Capitalist City*. Oxford: Basil Blackwell, 113–37.

Perry, D. and Watkins, A. 1977: *The Rise of the Sunbelt Cities*. Beverly Hills: Sage.

Persky, J. 1978: Dualism, capital-labor ratios, and the regions of the United States. *Journal of Regional Science*, 18:373–81.

Pfeffer, J. and Baron, J. 1988: Taking the workers back out: recent trends in the structuring of employment. In Staw, B. and Cummings, L. (eds) *Research in Organizational Behavior*, vol. 10. Greenwich, Conn: JAI Press, 257–303.

Phillips, A. 1971: *Technology and Market Structure: A Study of the Aircraft Industry*. Lexington: D.C. Heath.

Phillips A. and Taylor, B. 1980: Sex and skill: notes towards a feminist economics. *Feminist Review*, 6:79–88.

Piore, M. 1968: The impact of the labor market upon the design and selection of productive techniques within the manufacturing plant. *Quarterly Journal of Economics*, 92:602–20.

Piore, M. 1979: *Birds of Passage: Long Distance Migrants and Industrial Societies*. New York: Cambridge University Press.

Piore, M. 1987: Corporate reform in American manufacturing and the challenge to economic theory. Unpublished paper, Sloan School of Management, MIT.

Piore, M. and Sabel, C. 1984: *The Second Industrial Divide*. New York: Basic Books.

Pitman, B. 1987: Female designers and manufacturers in the early women's sportswear industry in Los Angeles. Unpublished paper, Graduate School of Urban Planning, University of California, Los Angeles.

Pollard, S. 1981: *Peaceful Conquest: The Industrialization of Europe, 1760–1970*. New York: Oxford University Press.

Polli, R. and Cook, V. 1969: Validity of the product life cycle. *Journal of Business*, 42:385–400.

Pomeroy, E. 1965: *The Pacific Slope*. New York: Albert Knopf.

Portes, A. and Walton, J. 1981: *Labor, Class and the International System*. New York: Academic Press.

Post, C. 1982: The American road to capitalism. *New Left Review*, 133:30–51.

Pratt, E. 1911: *The Industrial Causes of Congestion of Population in New York City*. New York: Columbia University Press.

Pred, A. 1966: *The Spatial Dynamics of Urban Growth in the United States, 1800–1914*. Cambridge, Mass: MIT Press.

Pred, A. 1974: *Urban Growth and the Circulation of Information, 1790–1840*. Cambridge, Mass: Harvard University Press.

Pred, A. 1977: *City-Systems in Advanced Economies*. London: Hutchinson.

Pred, A. 1980: *Urban Growth and City Systems in the United States, 1840–60*. Cambridge, Mass: Harvard University Press.

Preteceille, E. 1985: The industrial challenge and the French left. *International Journal of Urban and Regional Research*, 9(2):273–90.

Przeworski, A. 1985: *Capitalism and Social Democracy*. New York: Cambridge University Press.

Pudup, M. 1983: Packers and Reapers, Merchants and Manufacturers: Industrial Structuring and Location in an Era of Emergent Capitalism. Master's thesis, Department of Geography, University of California, Berkeley.

Pudup, M. 1987: From farm to factory: structuring and location of the US farm machinery industry. *Economic Geography*, 63(3):203–22.

Rainnie, A. 1984: Combined and uneven development in the clothing industry. *Capital and Class*, 22:141–56.

Rees, J. 1979: Technological change and regional shifts in American manufacturing. *The Professional Geographer*, 31(1):45–74.

Rees, J., Briggs, R. and Hicks, D. 1985: New technology in the US machinery industry. In Thwaites, A. and Oakey, R. (eds) *The Regional Economic Impact of Technological Change*. London: Frances Pinter, 164–94.

Reich, M. 1981: *Racial Inequality: A Political Economic Analysis*. Princeton: Princeton University Press.

Reich, R. 1986: *Tales of a New America*. New York: Times Books.

Ricardo, D. 1821: *The Principles of Political Economy and Taxation*. London: John Murray.

Richardson, H. 1973: *Regional Growth Theory*. London: Macmillan.

Robinson, J. 1956: *The Accumulation of Capital*. London: Macmillan.

Robinson, J. 1962: *Essays in the Theory of Economic Growth*. London: Macmillan.

Robinson, J. 1977: The labor theory of value. *Monthly Review*, 29(7):50–9.

Robinson, J. 1979: Foreword. In Eichner, A. (ed.) *A Guide to Post-Keynesian Economics*. White Plains, NY: Sharpe, xi–xxi.

Roemer, J. 1979: Divide and conquer: the micro-analytic foundations of the Marxian theory of wage discrimination. *Bell Journal of Economics*, 10(2):695–705.

Roemer, J. 1980: A general equilibrium approach to Marxian economics. *Econometrica*, 48(2):505–30.

Roemer, J. 1982: *A General Theory of Exploitation and Class*. New York: Cambridge University Press.

Rogers, E. and Larsen, J. 1984: *Silicon Valley Fever*. New York: Basic Books.

Rogers, E. and Schoemaker, F. 1971: *Communication of Innovations: A Cross-Cultural Approach*. New York: Free Press.

Rogers, J. 1985: Divide and conquer: the legal foundations of postwar US labor policy. Paper delivered at UCLA Political Economy Colloquium, December 12.

Rose, D. 1981. Accumulation versus reproduction in the inner city. In Dear, M. and Scott. A. (eds) *Urbanization and Urban Planning in Capitalist Societies*. London: Methuen, 339–82.

Rosen, K. and Retnick, M. 1980: The size distribution of cities, the Pareto law, and primate city size. *Journal of Urban Economics*, 8:165–86.

Rosenberg, N. 1972: *Technology and American Economic Growth*. New York: Harper Torchbooks.

Rosenberg, N. 1976: *Perspectives on Technology*. Cambridge, UK: Cambridge University Press.

Rosenberg, N. 1982: *Inside the Black Box: Technology and Economics*. Cambridge, UK: Cambridge University Press.

Rostow, W. 1961: *The Stages of Economic Growth*. Cambridge, UK: Cambridge University Press.

Rothwell, R. and Zegveld, W. 1985: *Reindustrialization and Technology*. London: Longman.

Rowthorn, R. 1975: What remains of Kaldor's laws? *Economic Journal*, 85:10–19.

Rubery, J. 1980: Structured labor markets, worker organization, and low pay. In Amsden, A. (ed.) *The Economics of Women and Work*. New York: St Martin's Press, 242–70.

Rubery, J. and Wilkinson, F. 1981: Outwork and segmented labor markets. In Wilkinson, F. (ed.) *The Dynamics of Labor Market Segmentation*. London: Academic Press, 115–32.

Russo, M. 1986: Technical change and the industrial district: the role of interfirm relations in the growth and transformation of ceramic tile production in Italy. *Research Policy*, 14:329–43.

Sabel, C. 1982: *Work and Politics*. New York: Cambridge University Press.

Sabel, C. and Zeitlin, J. 1985: Historical alternatives to mass production: politics, markets and technology in nineteenth century industrialization. *Past and Present*, 108:133–76.

Sahal, D. 1981: *Patterns of Technological Innovation*. Reading: Addison-Wesley.

Salter, W. 1966: *Productivity and Technical Change*. Cambridge, UK: Cambridge University Press.

Samuel, R. 1977: Workshop of the world: steam power and hand technology in mid-Victorian Britain. *History Workshop Journal*, 3:6–73.

Sassen, S. 1988: *The Mobility of Labor and Capital*. New York: Cambridge University Press.

Sato, K. 1987: Saving and investment. In Yamamura, K. and Yasuba, Y. (eds) *The Political Economy of Japan, Volume I: The Domestic Transformation*. Stanford: Stanford University Press, 137–85.

Sattlinger, J. 1975: *Capital and the Distribution of Labor Earnings*. Lexington, Mass: Lexington Books.

Saxenian, A. 1984: The urban contradictions of Silicon Valley. In Sawers, L. and Tabb, W. (eds) *Sunbelt-Frostbelt*. New York: Oxford University Press, 163–200.

Saxenian, A. 1985: Let them eat chips. *Society and Space*, 3:121–27.

Sayer, A. 1984: *Method in Social Science: A Realist Approach*. London: Hutchinson.

Sayer, A. 1985: Industry and space: a sympathetic critique of radical research. *Society and Space*, 3:3–30.

Sayer, A. 1986a: Industrial location on a world scale: the case of the semi-conductor industry. In Scott, A. and Storper, M. (eds) *Production, Work, Territory*. Boston: Allen and Unwin, 107–24.

Sayer, A. 1986b: New developments in manufacturing: the just-in-time system. *Capital and Class*, 30:43–72.

Schmenner, R. 1978: *The Manufacturing Location Decision*. Cambridge, Mass: Harvard Business School.

Schmookler, J. 1966: *Invention and Economic Growth*. Cambridge, Mass: Harvard University Press.

Schoenberger, E. 1986: Competition, competitive strategy and industrial change: the case of electronic components. *Economic Geography*, 62(4):321–33.

Schoenberger, E. 1987a: Technological and organizational change in automobile production: spatial implications. *Regional Studies*, 21(3):199–214.

Schoenberger, E. 1987b: From Fordism to flexible accumulation: technology, competitive strategies and international location. *Society and Space*, 6(3):245–62.

Schonberger, R. 1982: *Japanese Manufacturing Techniques*. New York: Free Press.

Schumpeter, J. 1934: *The Theory of Economic Development*. Cambridge, Mass: Harvard University Press.

Schumpeter, J. 1939: *Business Cycles*. New York: McGraw-Hill.

Schumpeter, J. 1942: *Capitalism, Socialism, and Democracy*. 1975 edition. New York: Harper and Row.

Scott, A. 1983: Industrial organization and the logic of intra-metropolitan location I: theoretical considerations. *Economic Geography*, 59:233–50.

Scott, A. 1985: Location processes, urbanization, and territorial development: an exploratory essay. *Environment and Planning A*, 17(4):479–503.

Scott, A. 1986: Industrialization and urbanization: a geographical agenda. *Annals of the Association of American Geographers*, 76(1):25–37.

Scott, A. 1987: The semiconductor industry in Southeast Asia: organization, location and the international division of labor. *Regional Studies*, 21(2):143–60.

Scott, A. 1988a: *Metropolis: From the Division of Labor to Urban Form*. Berkeley and Los Angeles: University of California Press.

Scott, A. 1988b: *New Industrial Spaces*. London: Pion.

Scott, A. and Angel, D. 1987: The global assembly operations of US semiconductor firms: a geographical analysis. *Environment and Planning A*, 19:875–912.

Scott, A. and Storper, M. 1987: High technology industry and regional development: a theoretical critique and reconstruction. *International Social Science Journal*, 1(12):215–32.

Scoville, J. 1973: *The Job Content of the U.S. Economy, 1940–1970*. New York: McGraw-Hill.

Scranton, P. 1983: *Proprietary Capitalism*. New York: Cambridge University Press.

Semmler, W. 1984: *Competition, Monopoly, and Differential Profit Rates*. New York: Columbia University Press.

Shaiken, H. 1984: *Work Transformed: Automation and Labor in the Computer Age*. New York: Holt, Reinhart and Winston.

Shaikh, A. 1978: The political economy of capitalism: notes on Dobb's theory of crisis. *Cambridge Journal of Economics*, 2(2):233–51.

Shaikh, A. 1980: Marxian competition versus perfect competition. *Cambridge Journal of Economics*, 4(1):75–83.

Shaikh, A. 1982: Neo-Ricardian economics: a wealth of algebra, a poverty of theory. *Review of Radical Political Economy*, 14(2):67–84.

Shaikh, A. 1984: The transformation from Marx to Sraffa. In Mandel, E. and Farjoun, E. (eds) *Ricardo, Marx, Sraffa*. London: Verso, 43–84.

Shapira, P. 1986: Industry and Jobs in Transition: A Study of Industrial Restructuring and Worker Displacement in California. Doctoral dissertation, Department of City and Regional Planning, University of California, Berkeley.

Shepherd, W. 1979: *The Economics of Industrial Organization*. Englewood Cliffs: Prentice-Hall.

Sheppard, E. 1982: City size distributions and spatial economic change. *International Regional Science Review*, 7:127–51.

Sheppard, E. and Barnes, T. 1986: Instabilities in the geography of capitalist production: collective versus individual profit maximization. *Annals of the Association of American Geographers*, 76(4):493–507.

Siegal, L. and Markoff, J. 1985: *The High Cost of High Tech*. New York: Harper and Row.

Singer, D. 1988: *Is Socialism Doomed? The Meaning of Mitterrand*. New York: Oxford University Press.

Singelmann, J. and Tienda, M. 1985: The process of occupational change in a service society: the case of the United States. In D. Gallie, (ed.) *New Approaches to Economic Life*. Manchester: Manchester University Press.

Singh, A. and Whittington, G. 1968: *Growth, Profitability and Valuation*. Cambridge, UK: Cambridge University Press.

Smith, A. 1776: *The Wealth of Nations*. 1937 edition. New York: Modern Library.

Smith, D. 1981: *Industrial Location*. 2nd ed. New York: Wiley.

Smith, M. and Feagin, J. 1987: *The Capitalist City*. Oxford: Basil Blackwell.

Smith, N. 1984: *Uneven Development*. Oxford: Basil Blackwell.

Sobel, R. 1983: *IBM: Colossus in Transition*. New York: Bantam Books.

Soja, E. 1980: The socio-spatial dialectic. *Annals of the Association of American Geographers*, 70(2):207–25.

Soja, E. 1986: Taking Los Angeles apart: some fragments of a critical human geography. *Society and Space*, 4(3):255–72.

Soja, E. 1988: *Post-Modern Geographies: The Reassertion of Space in Critical Social Theory*. London: Verso.

Solinas, G. 1982: Labor market segmentation and workers' careers: the case of the Italian knitwear industry. *Cambridge Journal of Economics*, 6:331–52.

Solow, R. 1957: Technical change and the aggregate production function. *Review of Economics and Statistics*, 39:312–20.

Solow, R. 1970: *Growth Theory*. New York: Oxford University Press.

Solow, R., Tobin, J., Von Weizacker, C. and Yaari, M. 1966: Neoclassical growth with fixed factor proportions. *Review of Economic Studies*, 33:79–116.

Spenner, K. 1983: Deciphering prometheus: temporal change in the skill level of work. *American Sociological Review*, 48:824–37.

Sraffa, P. 1926: The laws of returns under competition. *Economic Journal*, 36:535–51.

Sraffa, P. 1960: *Production of Commodities by Means of Commodities*. Cambridge, UK: Cambridge University Press.

Stanback, T., Bearse, P., Noyelle, T. and Kanasek, R. 1983: *Services: The New Economy*. Totowa, NJ: Allanheld, Osmun.

Starr, P. 1982: *The Social Transformation of American Medicine*. New York: Basic Books.

Steedman, I. 1977: *Marx After Sraffa*. London: New Left Books.

Steedman, I. and ten co-authors. 1981: *The Value Controversy*. London: Verso.

Steindl, J. 1952: *Maturity and Stagnation in American Capitalism*. Oxford: Basil Blackwell.

Stigler, G. 1951: The division of labor is limited by the extent of the market. *Journal of Political Economy*, 69:213–25.

Stigler, G. 1961: The economics of information. *Journal of Political Economy*, 69:213–25.

Stigler, G. 1962: Information in the labor market. *Journal of Political Economy*, 70:94–105.

Stone, K. 1981: The post-war paradigm in American labor relations. *Yale Law Journal*, 90:1509–80.

Storper, M. 1982: The Spatial Division of Labor: Technology, the Labor Process and the Location of Industries. Doctoral dissertation, Department of Geography, University of California, Berkeley.

Storper, M. 1984: Who benefits from industrial decentralization? Social power in the labor market, income distribution, and spatial policy in Brazil. *Regional Studies*, 18(2):143–64.

Storper, M. 1985: Oligopoly and the product cycle: essentialism in economic geography. *Economic Geography*, 61(3):260–82.

Storper, M. 1989: The transition to flexible specialization in industry. *Cambridge Journal of Economics*, 13:1–32.

Storper, M. and Christopherson, S. 1987: Flexible specialization and regional industrial agglomerations: the case of the US motion picture industry. *Annals of the Association of American Geographers*, 77(1):104–17.

Storper, M. and Scott, A. 1988: The geographical foundations and social regulation of flexible production complexes. In Wolch, J. and Dear, M. (eds) *The Power of Geography: How Territory Shapes Social Life*. Boston: Allen and Unwin, (forthcoming).

Strange, S. 1986: *Casino Capitalism*. Oxford: Basil Blackwell.

Sweezy, P. and Magdoff, H. 1988: *The Logic of Stagnation and the Financial Explosion*. New York: Monthly Review Press.

Sylos-Labini, V. 1962: *Oligopoly and Technical Progress*. English edition 1969. Cambridge, Mass: Harvard University Press.

Taylor, G. 1915: *Satellite Cities: A Study of Industrial Suburbs*. New York: Appleton.

Taylor, G. 1967: American urban growth preceding the railway age. *Journal of Economic History*, 27(3):309–39.

Taylor, M. 1975: Organizational growth, spatial interaction and location decision-making. *Regional Studies*, 9:313–23.

Taylor, M. and Thrift, N. (eds) 1982: *The Geography of Multinationals*. New York: St Martin's Press.

Teece, D. 1985: Multinational enterprise, internal governance and industrial organization. *American Economic Review, Papers and Proceedings*, 75(2):233–8.

Temin, P. 1966: Labor scarcity and the problem of American industrial efficiency in the 1850s. *Journal of Economic History*, 26:272–98.

Terleckyj, N. 1980: What do R and D numbers tell us about technological change? *American Economic Review*, 70(2):55–61.

Teulings, A. 1984: The internationalization squeeze: double capital movement and job transfer within Philips worldwide. *Environment and Planning A*, 16:597–614.

Therborn, G. 1984: The prospects of labour and the transformation of advanced capitalism. *New Left Review*, 145:5–38.

Thernstrom, S. 1964: *Poverty and Progress: Social Mobility in a Nineteenth Century City*. Cambridge, Mass: Harvard University Press.

Thirlwall, A. 1980: Rowthorn's interpretation of Verdoorn's laws. *Economic Journal*, 90:386–8.

Thomas, B. 1973: *Migration and Economic Growth*. Cambridge, UK: Cambridge University Press.

Thomas, M. 1975: Growth pole theory, technological change, and regional economic growth. *Papers of the Regional Science Association*, 34:3–26.

Thompson, W. 1968: Internal and external factors in the development of urban economies. In Perloff, H. and Wingo, L. (eds) *Issues in Urban Economics*. Baltimore: Johns Hopkins Press, 43–62.

Thurow, L. 1975: *Generating Inequality*. New York: Basic Books.

Thwaites, A. 1983: The employment implications of technological change in a regional context. In Gillespie, A. (ed.) *Technological Change and Regional Development*. London: Pion, 36–53.

Tiebout, C. 1954: The urban economic base reconsidered. *Land Economics*, 32:95–9.

Timms, D. 1971: *The Urban Mosaic*. Cambridge, UK: Cambridge University Press.

Truel, J. 1980a: *L'Industrie Mondiale des Semiconducteurs*. Dissertation doctorale de 3e cycle en Sciences Économiques, Université de Paris, Dauphine.

Truel, J. 1980b: Les nouvelles stratégies de localization internationale: le cas des semi-conducteurs. *Revue d'Économie Industrielle*, 14(4):171–78.

Truel, J. 1983: Structuration en filière et politique industrielle dans l'électronique: une comparison internationale. *Revue d'Économie Industrielle*, 23(1):293–303.

Turner, F. 1920: *The Frontier in American History*. New York: Holt.

United States Bureau of Economic Analysis. 1978: *Occupational Pay Comparisons*. Washington, DC: US Government Printing Office.

Van Duijn, J. 1983: *The Long Wave in Economic Life*. London: Allen and Unwin.

Vance, J. 1977: *This Scene of Man*. New York: Harper and Row.

Vance, J. 1986: *Capturing the Horizon: The Historical Geography of Transportation Since the Transportation Revolution of the Sixteenth Century*. New York: Harper and Row.

Vennin, R. and De Banville, E. 1975: Pratique et signification de la sous-traitance dans l'industrie de l'automobile en France. *Revue Économique*, 26(2):280–306.

Verdoorn, P. 1980: Verdoorn's law in retrospect: a comment. *Economic Journal*, 90:382–85.

Verlaque, E. 1984: Trente ans de décentralisation industrielle en France, 1954–1984. In Bastié, J. (ed.) *Trente ans de décentralisation industrielle en France*. Cahiers No. 7. Paris: Centre de Recherches et d'Études sur Paris et l'Île-de-France, 7–182.

Vernon, R. 1960: *Metropolis 1985: An Interpretation of the Findings of the New York Metropolitan Region Study*. Cambridge, Mass: Harvard University Press.

Vernon, R. 1966: International investment and international trade in the product cycle. *Quarterly Journal of Economics*, 80:190–207.

Vietorisz, T. and Harrison, B. 1973: Labor market segmentation: positive feedback and divergent development. *American Economic Review*, 63:366–76.

Wachter, M. and Williamson, O. 1978: Obligational markets and the mechanics of inflation. *Bell Journal of Economics*, 9:549–57.

Walby, S. 1986: *Patriarchy at Work*. Oxford: Polity Press.

Walker, P. 1979: *Between Labor and Capital*. Boston: South End Press.

Walker, R. 1977: The Suburban Solution: Capitalism and the Construction of Urban Space in the United States. Doctoral dissertation, Department of Geography and Environmental Engineering, Johns Hopkins University, Baltimore.

Walker, R. 1981: A theory of suburbanization: capitalism and the construction of urban space in the United States. In Dear, M. and Scott, A. (eds) *Urbanization and Urban Planning in Capitalist Societies*. New York: Methuen, 383–430.

Walker, R. 1985a: Is there a service economy? The changing capitalist division of labor. *Science and Society*, 49:42–83.

Walker, R. 1985b: Class, division of labor and employment in space. In Gregory, D. and Urry, J. (eds) *Social Structure and Spatial Relations*. London: Macmillan, 164–89.

Walker, R. 1988a: The dynamics of value, price and profit. *Capital and Class*, 35:146–81.

Walker, R. 1988b: Machinery, labor and location. In Wood, S. (ed.) *The Transformation of Work?* London: Hutchinson (forthcoming).

Walker, R. 1988c: The geographical organization of production systems. *Society and Space*, 7:377–408.

Walkowitz, D. 1978: *Worker City, Company Town*. Champaign: University of Illinois Press.

Wallerstein, I. 1974: *The Modern World-System*. New York: Academic Press.

Warntz, W. 1965: *Macrogeography and Income Traits*. Philadelphia: Regional Science Research Institute, Monograph Series No. 3.

Warren, R. 1978: *The Community in America*. Chicago: Rand McNally.

Watts, H. 1980: *The Large Industrial Enterprise: Some Spatial Perspectives*. London: Croom-Helm.

Watts, H. 1981: *The Branch Plant Economy*. London: Longman.

Webber, M. 1964: The urban place and the non–place urban realm. In Webber, M. (ed.) *Explorations into Urban Structure*. Philadelphia: University of Pennsylvania Press, 79–153.

Webber, M. J. 1987: Rates of profit and interregional flows of capital. *Annals of the Association of American Geographers*, 77(1):63–75.

Webber, M. J. 1988: Profits, crises and industrial change: Canada, 1952–81. *Antipode*, 20:1–32.

Weber, A. 1909: *Theory of the Location of Industries*. English edition 1929. Chicago: University of Chicago Press.

Weeks, J. 1981: *Capital and Exploitation*. Princeton: Princeton University Press.

Wells, L. (ed.) 1972: *The Product Life Cycle and International Trade*. Boston: Harvard Business School.

Wheeler, J. and Brown, C. 1985: The metropolitan corporate hierarchy in the US South, 1960–1980. *Economic Geography*, 61(1):66–78.

Wilkinson, F. (ed.) 1981: *The Dynamics of Labor Market Segmentation*. London: Academic Press.

Williams, K., Williams, J. and Thomas, D. 1983: *Why are the British Bad at Manufacturing?* London: Routledge and Kegan Paul.

Williamson, O. 1975. *Markets and Hierarchies*. New York: Free Press.

Wolf, E. 1982. *Europe and the People Without History*. Berkeley: University of California Press.

Wollenberg, C. 1988: *Marinship: The Rise and Fall of a Military-Industrial Complex*. Unpublished book manuscript, Vista College, Berkeley.

Wood, P. 1986: *Southern Capitalism: The Political Economy of North Carolina*. Durham: Duke University Press.

Wood, S. 1982: *The Degradation of Work?* London: Hutchinson.

Wood, S. 1988: *The Transformation of Work?* London: Hutchinson.

Woodward, J. 1969: *Industrial Organization, Behavior, and Control*. London: Oxford University Press.

Wright, E. 1976: Class boundaries in advanced capitalist societies. *New Left Review*, 98:3–41.

Wright, E. 1985: *Classes*. London: Verso.

Young, A. 1928: Increasing returns and economic progress. *Economic Journal*, 38:527–42.

Young, R. 1986: Industrial location and regional change: the United States and New York state. *Regional Studies*, 20(4):341–69.

Zeitlin, J. 1979: Craft control and the division of labor: engineers and compositors in Britain 1890–1930. *Cambridge Journal of Economics*, 3:263–74.

Zimbalist, A. 1979: *Case Studies on the Labor Process*. New York: Monthly Review Press.

Zipf, G. 1949: *Human Behavior and The Principle of Least Effort*. New York: Hafner.

Zunz, O. 1982: *The Changing Face of Inequality: Urbanization, Industrialization and Immigration in Detroit, 1880–1920*. Chicago: University of Chicago Press.

Zysman, J. 1977: *Political Strategies for Industrial Order*. Berkeley: University of California Press.

Names Index

Subject Index